Poultry Science, Chicken Culture

Poultry Science, Chicken Culture

 artial  lphabet

SUSAN MERRILL SQUIER

RUTGERS UNIVERSITY PRESS

NEW BRUNSWICK, NEW JERSEY, AND LONDON

LIBRARY OF CONGRESS CATALOGING-IN-PUBLICATION DATA

Squier, Susan Merrill.
    Poultry science, chicken culture : a partial alphabet / Susan Merrill Squier.
        p.   cm.
    Includes bibliographical references and index.
    ISBN 978-0-8135-4924-8 (hardcover : alk. paper)
    1. Chickens.   2. Chickens—Social aspects.   3. Animal culture—Moral and ethical
aspects.   I. Title.
    SF487.S74   2011
    636.5—dc22

                                                                2010013765

A British Cataloging-in-Publication record for this book
is available from the British Library.

Visit our Web site: http://rutgerspress.rutgers.edu

Manufactured in the United States of America

For Connie, who taught me to love chickens,
and for Gowen, who has learned to love them

CONTENTS

ACKNOWLEDGMENTS

I have far too many people to thank to name them all, but let me make a beginning in the hope that those I neglect to mention will nonetheless accept my sincere gratitude. I thank the following places for inviting me to give talks based on early versions of chapters: the "Finding Animals" conference sponsored by the "Visualizing Animals" group at Penn State University, where I tried out "Hybridity"; York College, in York, Pennsylvania, whose Honors Program was a friendly host for early versions of "Biology" and "Inauguration"; the "Performativities Conference" sponsored by the Gender Institute at the London School of Economics, where I also presented a version of "Biology"; the Medical Museum at the University of Copenhagen and the Platform in Life-Science Governance at the University of Vienna, where I presented parts of "Disability"; the "Science Futures" conference of the Swiss STS, where I tried out a portion of "Hybridity"; the Cultural Studies Association Annual Conference in Portland, Oregon, where I gave a portion of "Why Chickens?" as well as an early version of "Fellow-Feeling"; the Amsterdam meeting of the European Conference for the Society for Literature, Science, and the Arts, where I presented a version of "Auguries"; the conference "Who Owns Knowledge? A Symposium on Science and Technology in the Global Circuit" sponsored by George Mason University, where I tried out "Poultry Science, Chicken Culture"; and the Society for Literature, Science, and the Arts, where I presented "A Manifesto for Agricultural Studies," the first talk of this nascent research project. Once again I thank members of the Society for Literature, Science, and the Arts for being such great interlocutors and friends. Earlier versions of several chapters of this book appeared in the following publications: "Chicken Auguries," *Configurations* 14 (2006): 69–86; "Liminal Livestock," *Signs* 35, no. 2 (Winter 2010): 477–502; "The Sky Is Falling: Risk, Safety, and the Avian Flu" in *The Rhetoric of Safety*, ed. Lawrence R. Schehr, Special Issue, *South Atlantic Quarterly* 107, no. 2 (Spring 2008): 387–410; and "Fellow-Feeling," in *Animal Encounters*, ed. Tom Tyler and Manuela Rossini, 173–196 (Leiden: Brill, 2009).

Sandra Stelts, curator of Rare Books and Manuscripts at Paterno Library of Penn State University, is top of my list to thank. One day I must have mentioned to her that I was working on a book on chickens. Imagine my surprise and gratitude when I received a package in the mail: a list of all the archival materials relating to chickens held in our superb Special Collections Library. Sandy has continued to keep her eyes open for chicken materials throughout this project, and it is she who first introduced me to Miss Nancy Luce, to Elmer Boyd Smith's beautiful children's book, *Chicken World*, and to many other early twentieth-century gems preaching the joys of the hen.

I am grateful to the members of PASA, the Pennsylvania Association for Sustainable Agriculture, for introducing me not only to the concept of pastured poultry but also to the wonderful Eli Reiff. I thank the following people for the generosity they showed in giving me interviews, valuable information, even guided tours: Johannes Paul and the other partners in Omlet for a wonderful visit and an inspiring story; Professor Claudio Stern of University College, London, for being so generous with his time and information; Paul Farley of Seattle Tilth, the first person I interviewed in the course of this long project, for sharing with me his copy of E. B. White's "Introduction"; Faith Wilding, for sending me subRosa's great pamphlet on women and chickens; Jean Pagliuso, for her willingness to share experiences and photographs with me when I called on her after viewing her "Poultry Suite" show at the Marlborough Gallery; Koen Vanmechelen, for his hospitality, and for being a kindred spirit and a mind-expanding artist; Ruth Ozeki, for a writing collaboration that blossomed into a shared love of chickens and Zen meditation; Sandi Morgen, Nancy Tuana, and the Rock Ethics Institute at Penn State for the faculty seminar on Social Justice and the Economy that got me thinking about the economics of chickens; Lovalerie King for her helpful comments in that seminar, and for her terrific book on race and theft; Grant Farred, for inviting me to write the first draft of "Epidemic"; my young neighbor Sam for splitting a flock of six Buff Wyandottes we bought at auction at the Belleville Poultry Auction; Patricia Dunn, DVM, of the Animal Diagnostic Lab, professors Ramesh Ramachandran and Robert Elkins of the Poultry Science Program, and poultry educator Philip Clauer of the Program in Agricultural Extension Education, all of Penn State University, for patiently sharing their knowledge and expertise with such a rank beginner in the chicken business.

I thank the Writer's Room in New York City for being a zone of relaxed intensity in which to write for a sabbatical semester; the librarians at the National Agricultural Library, in Beltsville, Maryland, for their hospitality even as their space and support had been cut dramatically; Professor Joy Pate, director of Penn State's Center for Reproductive Biology and Health, for a stimulating experiment in cross-college collaboration when students in my doctoral seminar on "Gender and Science: Reproduction" visited researchers at the center. I thank the dean of the College of Liberal Arts at Penn State, Professor Susan Welch, and the associate dean, Raymond Lombra, for financial support for this book's research and publication. I thank professors Carolyn Sachs, head of

Women's Studies, and Robert Caserio and Robin Schulze, heads of the English department, for being willing to suspend disciplinary categories—indeed, to suspend disbelief—as I worked on this project. I thank Anne Buchanan for those meetings over coffee to talk about goats and chickens, and for answering my many questions about biology and genetics; Clare Hinrichs, for all those conversations about agriculture and rural sociology; Janet Lyon, for giving me my subtitle, and, with Michael Bérubé, for creating a lively disability studies community at Penn State; and Irina Aristarkhova, for sharing with me the slow process of writing, rethinking, and rewriting. As always, I thank my graduate students, Megan Brown, Melissa Littlefield, Shannon Walters, Jenell Johnson, Sarah Birge, and Angela Ward, for assistance with research; and the members of my doctoral seminars in English and Women's Studies for their illuminating comments, suggestions, and responses.

I thank Dorn, Gabeba, Ted, Pat, Leif, and Gowen for sitting with me, and Marie for walking with me, through years of chicken obsession and writing; I thank Jennifer for sharing the enthusiasm; and I thank my sister Virginia and her family, and Gowen, Caitlin, and Toby, for tolerating conversations about chickens that went on and on and on. (You did seem to enjoy those delicious eggs, though!) My father, John Squier, introduced me to chickens in my childhood. He kept a flock of ornamental birds—I remember Barred Rocks, Polish, Houdans, Brahmas, and Plymouth Rocks—until the chicken coop leached into the well beneath it, and he got hepatitis. I thank my sister Robin for first-rate birding lore.

I thank my mother, Connie Squier, for setting such a good example. She let a rooster with frostbitten feet live in our basement for one entire winter.

Despite all the help I received, there are certain to be errors and omissions—sadly, those are mine alone.

Poultry Science, Chicken Culture

Introduction

Why Chickens?

"Don't try to convey your enthusiasm for chickens to anyone else."

– E. B. White, Introduction to Roy E. Jones's
A Basic Chicken Guide for the Small Flock Owner

The Chickens and I

One morning in July, several years ago, I got a phone call from a man who identified himself as Glenn, from the Poultry Education and Research Center (PERC) at my university. "When do you want to pick up your hens?" he asked. Several weeks earlier, I had been searching for chickens to build up my flock. My favorite hen had died, and I had realized I should call PERC. After all, I had long rejoiced in working at what I called a "full-service university": it was time to draw on some of those services. And so, I had called Glenn to ask about chickens.

Did they have any laying hens, I had asked. Nope, they had meat birds, he told me. I was pretty dubious about using a meat bird as a laying hen. But we had left it that he would call me when he had some hens that were old enough for me to look at them, to decide whether I would add them to my flock. Now whatever birds they had were apparently old enough for that look.

"What breed are they?" I asked.

"Cobbs."

"But that's not a breed, that's a brand," I replied. "What breed are they?"

I had done a little bit of research into chickens by then, and thought I was pretty on top of things. I had discovered that Cobbs, like other meat birds raised by agricultural schools in their poultry science programs, were hybrids created by poultry breeding companies specifically for the demands of the poultry market. This was pretty interesting to someone who teaches and does research in science studies for a living. Glenn explained, patiently, that Cobbs were created by crossing two breeds of chickens: the Cornish and the Barred Rock. They were the standard hybrid produced as broilers by the poultry industry. Would they work for laying hens? I wondered. He figured they could prove adequate layers if I allowed them free range and didn't give them food on demand. (That would make them bulk up and stop laying, he warned me.)

These particular birds had been raised in a flock scheduled to be butchered at six weeks, he told me. They were one day short of four weeks old today. Did I want them? Scheduled to die in two weeks. Two weeks during which they would be closed in small breeding cages, offered food nonstop, and kept under lights twenty-three of every twenty-four hours. I told him I would be over in an hour. I was ready to move fast not because I was committed to poultry rescue—though that is a growing part of chicken culture—nor because I was increasingly interested in the ethical and political implications of our ways of growing food but simply because after several years of keeping chickens, I'd learned that "started pullets"—chickens not yet mature but no longer needing incubation—are very hard to find even in rural central Pennsylvania.

When Glenn called, I was down to just one hen, a little bantam who was laying one egg almost every day. Two other hens had come and gone as layers. A gorgeous, pheasant-like brown and golden hen with splendid silky leggings had died while I was out of town. And a competent Buff Orpington—named Lazarus because she rose from seeming death after being chased by my Whippet—finally succumbed when a raccoon got her. When she died, I stopped naming my chickens, but I still hoped for a larger flock and more eggs. I had plans to go to the live bird auction at the Amish market one valley south on the very next day. But now I had Glenn on the phone. Two scientifically raised birds in the hand are worth more than any number of proverbial birds in the next valley, I decided. So off I went to pick up the two little Cobb pullets.

I met Glenn at the white barns of the Poultry Education Research Building on the outskirts of campus. He handed me a box labeled "Eggs: The Incredible Edible Egg." I peered inside to find two scrawny, sparsely feathered white pullets, heavy chests like bony keels, and thick, sturdy yellow legs startlingly outsized for their welterweight heft. These weren't just Cobbs. They were Cobb 700s, I later learned when I checked the Cobb-Vantress, Inc., Web site. They were more like an LX model than the standard DL. That standard model was the Cobb 500, advertised as "The World's Most Efficient" chicken because it featured the "lowest feed conversion together with the ability . . . to thrive on lower density, less expensive feed, [which] reduces the cost of producing chicken meat, [decreasing] feed ingredient costs without reducing performance."[1] This newer model, the Cobb 700, was developed—so the Web site explained—to meet the demand for even more breast meat. Aimed at the value-added processed chicken markets, it enabled "customers [to] optimize grow out and processing performance, achieving the highest eviscerated and breast meat yield of all breed options."[2] *Some Bird*, I thought: not only able to grow 1 percent more breast meat than its nearest rival (sibling? floor model?) but to maximize the profits from its own processing! And all this in a bird that did all its grow-ing between four and six weeks, at which point, having attained its target weight, it was slaughtered to produce the "best overall feed conversion" and "best breast meat conversion" that made the Cobb-Vantress corporation so proud of its product.[3]

I hefted the box with the two little pale pullets into the car—not much work there, since they were still pretty scrawny despite the weight gain promised in the next two weeks—and took them home to introduce them to the bantam. Traditional British mixed breed, meet American hybrid brand of the future. Nature, meet culture. So shy and weak that they let me pick them up, they flopped into the shavings where I dropped them in the chicken shed. Their out-sized legs did not yet seem able to carry them, or maybe it was that they had spent their entire four weeks of life on a wire floor, and had no idea how to nav-igate the textures and topography of a shavings-filled henhouse. In any case, they were a sorry-looking pair. I walked away from the henhouse thinking hard about their abject place in the new pecking order as well as my role as a novice chicken keeper. Where did we both stand in relation to the academic field of poultry science, the cultural sphere in which chickens were icons of domesticity and sources of comedy, and the global poultry industry that produced and shaped them as consumer objects? Over the years to come, those thoughts developed into this book.

Still: why chickens? Admittedly, the subject might seem an unlikely one for a teacher of women's studies and English. "Talking about chickens is a risky thing," wrote E. B. White more than half a century ago, in a little-known essay on chicken keeping. I am a long-time admirer of E. B. White—author of *Char-lotte's Web* and *Stuart Little* as well as the influential prose handbook *Elements of Style* (1962)—but until I began this book I'd never known that White was also a chicken farmer, keeping a flock on his Connecticut acres. In fact, he con-tributed the introduction to a small volume published by extension poultry-man, Roy E. Jones, *A Basic Chicken Guide for the Small Flock Owner* (1944).[4] White captures the pleasures of chickens as well as the dangers in this gem of an essay. Admitting an "attachment to the hen that dates from 1907," White acknowl-edges that not everyone shares his enthusiasm. To most of his city friends, he explains, the hen is merely "a comic prop straight out of vaudeville. When I would return to city haunts for a visit, these friends would greet me with a patronizing little smile and the withering question: 'How are all the chickens?'" (White 1944, v).

White wrote in 1944, at a time when the demand for chicken meat and eggs to supplement a protein-meager wartime diet had suddenly made chicken raising fashionable, and the volume he introduces is clearly a response to those economic and political realities. Jones's *Basic Chicken Guide* is direct, practical, and prosaic; on the front jacket flap it makes its case: "*thinking of keeping chickens?* Meat rationing and egg shortages won't bother you then." In chapters that proceed from "Should I Keep Poultry" and "When and How to Start" to "Growing Meat Chickens," "Poultry Breeding," and "Basic Ways to Cook Poultry and Eggs," poultryman Jones takes for granted an interest in chicken keeping motivated by wartime food rationing and economic stringency, providing a how-to manual that will guide even the most rank novice.

White's introduction to the guide transforms the subject of chicken keeping. Suggesting some of the elastic appeal of the chicken as subject, his essay mingles different kinds of knowledge and different voices in an exuberant embrace of the position of beginner: "Since this book is a guide, I feel I should instruct the reader, and should not only praise the hen but bury her. Luckily I can squeeze everything I know about chickens into a single paragraph. . . . Here, then, is my Basic Chicken Guide." What follows reads like a prose poem: "Be tidy. Be brave. Elevate all laying house feeders and waterers twenty-two inches off the floor. Use U-shaped rather than V-shaped feeders, fill them half full, and don't refill till they are empty. Walk, don't run. Never carry a strange object into the henhouse with you. Don't try to convey your enthusiasm for chickens to anyone else" (White 1994, vi). White then follows up this genuinely useful advice on chicken keeping with an almost Wordsworthian tribute to the practice. "The feeling I had as a boy for the miracle of incubation, my respect for the strange calm of broodiness, and my awe at an egg pipped from within after twenty-one days of meditation and prayer—these have diminished but slightly" (viii).

As I read White's essay, I mentally noted his good suggestions that waterers be elevated (so the chickens don't fill them with shavings) and feeders not be refilled until completely empty. But most of all, I marveled at the way the essay switched easily between practicality and poetry. White's essay exemplifies something I have found to be true: chickens are good to think with. For people in a wide range of fields, chickens have long provided a useful, and even at times inspiring, focus. Not just farmers (male and female) and farm wives but economists, politicians, historians, biologists, and doctors as well as artists of all kinds have found the chicken weaving its way into their thoughts, dreams, and deeds, shaking up the tidy separation of nature and culture and prompting exploration of what it means to be human—*and animal.*

As White himself acknowledged, it helps to be comfortable with humility if you are trying to persuade people who are not farmers that it is worth thinking about chickens. Yet, to be a humble novice does not necessarily mean to simplify. The generous genre used by White and Jones, the ABC or *abecedarium*, has helped many writers open up specialized subjects to broad audiences. Two other examples of the ABC also influenced me. Raymond Williams, founding theorist of British cultural studies, used the genre to organize his brilliant *Keywords: A Vocabulary of Culture and Society.* The table of contents reveals the surprising potency of the resulting juxtapositions, as it moves from

A
Aesthetic
Alienation
Anarchism
Anthropology
Art

to

W
Wealth
Welfare
Western
Work   (Williams 1983)

This little marvel, the ABC form, is also elastic enough to suit a postmodern philosopher and Marxist literary critic. Interviewer Claire Parnet used it for her conversation with her former teacher Gilles Deleuze, providing free range for the philosopher's far-flung intellectual interests. *L'Abécédaire de Gilles Deleuze* presents Deleuze discoursing on twenty-six alphabetically arranged themes in a serendipitous mixture of intimate anecdotes and philosophical explorations (Boutang and Parmat 1988). Take, for example, this paraphrase of an excerpt from his first entry, "Animal":

> What is essential, claims Deleuze, is to have an animal relationship with animals. Deleuze draws his conclusions from watching people walking their dogs down his isolated street, observing them talking to their dogs in a way that he considers "frightening" [effarant]. He reproaches psychoanalysis for turning animal images into mere symbols of family members, as in dream interpretation. Deleuze concludes by asking what relation one should or could have with an animal and speculates that it would be better to have an animal relation (not a human one) with an animal. Even hunters have this kind of relation with their prey.
>
> (Stivale 2007)

Charmed by these earlier precedents, and mindful of my beginner status in chicken raising, I found myself drawn to the ABC genre as this book began to take shape. I enjoyed its seemingly haphazard alphabetical organization, which helped me to escape disciplinary boundaries, perhaps even to approach the chicken on its own terms and forge what Deleuze described as "an animal relation with animals" (Stivale 2007). And although when I accepted those first

Cobb 700 pullets from Glenn at PERC I had planned to focus on the rise of an alternative to industrial chicken farming, I found myself wandering under the influence of the ABC form beyond that realm to the many roles chickens have played in knowledge production. In the process, I found myself documenting the production, in "fact" and fiction, of a new sort of "chicken culture."

Taking chickens as an augury—an omen, token, or indication—of the interwoven state of biology and culture in contemporary America, I have followed the thread of my own curiosity to assemble an alphabetical series of ten case studies of (poultry) science as (chicken) culture. Moving from augury, biology, and culture through disability, epidemic, and fellow-feeling to gender, hybridity, and inauguration, I explore the unanticipated effects of industrial rationality; the gendered, raced, and classed nature of agricultural as well as cultural production; and the varieties of knowledge produced by the human encounter with the chicken.

A Manifesto for Agricultural Studies

It will be clear by now that I have written this book in conversation with the fields of science studies, environmental studies, and animal studies. I acknowledge the immense contribution of modern industrial agriculture in feeding the rapidly increasing global population, but I share the belief of the authors of a recent article in *Nature* that such recent industrial agricultural practices have had "inadvertent, detrimental impacts on the environment and on ecosystem services" (Tilman et al. 2002, 672). In particular, I share their position that high-density, large-scale industrial animal farming "[has] health and environmental costs that must be better quantified to assess their potential role in sustainable agriculture" (674). I go beyond the authors' argument, however, because I believe that such costs must not merely be *better quantified*—they must be understood and expressed more fully, which means moving beyond merely positivist modes of knowing. I am convinced, after years of research and conversation with scholars in the fields of literature, women's studies, and science studies, that such a broader understanding and fuller mode of expression of what we understand can help us find ways of growing food that are both sustainable and equitable.

With this study of chickens, and agriculture more broadly, I return to two themes from my earlier work on biomedicine: my interest in the ways that human biomedicine has used other species as what I call research reservoirs and my commitment to a transdisciplinary mode of inquiry (Squier 2004). As I did earlier in *Liminal Lives*, once again I explore the social and scientific effects of the mining of female life—now both human and avian, in the agricultural as well as the medical sciences—for intellectual lore and economic ore. And I work at and between the boundaries of the disciplines because I am convinced that the resulting strategic marginality affords me perspectives unavailable to those wearing strictly disciplinary lenses.

I depart from my earlier writings in the explicit call for agricultural studies: an alternative, or at the very least a supplement, to traditional cultural studies (though the phrase itself may seem like a tautology). Cultural studies have shown little appreciation of the agricultural issues that Raymond Williams evoked so memorably in his relatively late work, *The Country and the City* (Williams 1973).[5] Instead, the field has typically taken a markedly metropolitan approach to the exploration of relations between individuals and societies, focusing primarily on cultural practices springing from industrialization, urbanization, and globalization. Growing from Williams's earlier explorations of the "Industrial Revolution, and of its consequent social and political changes" (Williams 1989, 10) as well as from the work of Richard Hoggart, the field of cultural studies has emphasized engaged analysis, a focus on subjectivity, attention to the relations between culture and individual lives, and a commitment to the investigation of the impact of the political and technological centralization of first world power in the great cities of the global North (During 1993; Hoggart 1998; and Williams 1989). We can see this emphasis clearly if we sample just one (well-known) text from the field, Simon During's *The Cultural Studies Reader* (1993). Although it includes articles and excerpts on nationalism, postcolonialism, and globalization; on ethnicity and multiculturalism; on science and cyberculture; on sexuality and gender; on carnival and utopia; on consumption and the market; on leisure; and on culture, it makes no reference at all to the practice of cultivating plants or animals for food (1993, 1, 2). This agricultural amnesia is curious when we consider that the word "culture" derives from the Latin word *cultura* and as far back as 1420 designated "the cultivation of a plant or crop" (OED 1971, 1247).

Agriculture should be of central importance to cultural studies because, as an institution and a set of practices, agriculture shapes our bodies, selves, societies, landscapes, geopolitical possibilities, and even the other species with which we share the Earth. The term *agriculture* has a broad reach, of course, from Stone Age plows to the pastoral. Although cultural studies may have given agriculture short shrift, discussions of rural and agrarian life have long animated literary history and criticism (Alpers 1997; and Conlogue 2001, 6). The pastoral has structured English and American literature. It has given us the motif of a blessed retreat from the pressures of an urban world, "a green thought in a green shade" as well as what D. H. Lawrence has described as a classically American escape from European cultural dominance (Marvell 1938). Indeed, agriculture has stirred partisan rivalries in the American landscape. So, twelve Southerners known as the "Agrarians" affirmed, in their 1930 introductory essay to the classic manifesto *I'll Take My Stand*, that they all "tend to support a Southern way of life against what may be called the American or prevailing way; and all . . . agree that the best terms in which to represent the distinction are contained in the phrase, Agrarian *versus* Industrial" (Lawrence 1964). Henry Nash Smith explores the myth of the garden as a foundational belief in American culture and politics, Leslie Fiedler argues that American men choose the wilderness because they associate it with masculine freedom from the civilizing domesticities of a feminized town life, and both Leo Marx and Annette Kolodny associate the rural with the feminine, whether that link invites consolation and embrace or leads to violence and exploitation (Conlogue 2001, 6; Fiedler 1992; Kolodny 1975; Marx 1974; and Smith 1970).

In contrast to such literary and historical explorations of the pastoral, the agricultural studies perspective I advocate can best be described as "postpastoral." I borrow this term from Terry Gifford, who frames its defining vision as awe at the natural world, recognition of a universe that is both creative and destructive, realization that inner and outer nature must be understood in relation to each other and that nature and culture are coextensive, acknowledgment that "with consciousness comes conscience," and an ecofeminist understanding that exploitation of women and minorities emerges from the same state of mind as environmental exploitation (Gifford 2000, 220).

We can further clarify this concept of the postpastoral by considering the contemporary response to *I'll Take My Stand*. Recent scholars have located populism, anti-industrialism, cultural conservatism, and whiteness in that volume's celebration of rural heritage (Donaldson 2006, ix). For example, Tanya Ann Kennedy argues that the volume formulates a gendered relation between private and public worlds that leads to a suppression of women writers even within a reclaimed regional identity. Kennedy charges the Agrarians with inconsistency and gender bias, arguing that despite their attention to the "displacements and alienations engendered by imperialist, industrial, and urban impositions upon a primarily agricultural people," they fail to take into account the differences between male and female agricultural work. Instead, their defense of the agrarian way of life "subsumes all white female members under the rubric of the household economy," shoving women to the margin "as an agent of the consumerism and sexuality associated with northern urbanism" (Kennedy 2006, 45).

"Postpastoral" is also associated with a paradigm shift in agriculture, as the general concept of farming changed from a way of life to a business, subject to the same strategies of rationalization, management practices, and control technologies as other industrialized businesses. Of course, this broad claim omits the various niches within industrial farming where small, marginal, and recreational farmers have continued to exist alongside the rationalized and large-scale agricultural holdings on the high plains at the turn of the twentieth-first century. It also relies on a restrictive and futuristic view of technology. Even the "Bonanza" wheat farms and the vast cattle ranches were forged by such low-technology tools as the mule-drawn plow and the barbed wire fence.[6] Yet while scholars will date the emergence of postparadigm agriculture earlier or later depending on the types of agricultural practices they investigate, a major reference point is arguably England from 1649 to 1650, when a group of Agrarian reformers known as the Diggers launched a "rural, radical and short-lived response to the enclosure laws and the widespread poverty and starvation" they had produced (Lyon 1999, 17).

The Diggers' argument—that poor people were a distinct group with a shared interest in access to the common lands that had recently been fenced in for the land-owning ruling class by the Enclosure Acts—was legitimated by religious understanding. They drew on the thinking of reformers Robert Coster and Gerard Winstanley, who held that the Earth was a "common treasury" bestowed by God for all men and women to use. While the Digger uprising was quickly crushed by the land-owning classes, their communities shattered, houses burned, and property destroyed, the rhetorical innovation they displayed remains. "Digger texts (and specifically Winstanley's writings) principally aimed to 'win over a rural proletariat (and other sympathetic groups) to a programme of mass political action'" (Lyon 1999, 19). They called, in songs and poems, "Stand up now, Diggers all," and the group whose solidarity they invoked was both called into being and given a cause. And so was launched a new literary genre, the manifesto.

This new genre embodies the troubled connections between modernity and agriculture, the ability to work one's own land and hold the status of citizen. A central theme of the manifesto is the protest that "We" (its empowered and oppressed collectivity) have not shared the fruits of modernity's technological and political innovations. Yet the manifesto's call for access to technological progress clashes with its origins in an Agrarian protest; the Digger movement waged a bitter and ultimately doomed fight against a new technology of land management: the fencing in, or enclosure, of common lands. "Historically," Janet Lyon reminds us, "manifestoes . . . appear most often in clusters around those crises that involve definitions of citizenship and political subjecthood" (Lyon 1999, 16). The very form of the manifesto performs the creation of a community, a "we" articulating its claim to the fruits of progress produced by contemporary science and medicine. Yet this new community also marked the end of an old type of individual: the subsistence farmer who worked lands held in common.

The initial act of agricultural dispossession not only led to the manifesto, but it also shaped the political subject: the functional, normative body of the citizen. We can see this notion of the citizen taking shape in John Locke's *Two Treatises of Government*, where Locke makes a distinctly agricultural case for the ownership of property: "*As much Land* as a Man Tills, Plants, Improves, Cultivates, and can use the Product of, so much is his *Property*. He by his Labor does, as it were, inclose it from the Common. . . . He that in Obedience to this Command of God, subdued, tilled and sowed any part of it, thereby annexed to it something that was his *Property*, which another had no Title to, nor could without injury take from him" (Locke 1689, 290–291). Locke's accomplishment lay not merely in forwarding a new notion that labor is the warrant for property, or in linking that property to a man's body, but in implicitly restricting the kinds of bodies that can produce property. He saw the proper body as a commodity, which, although it established a person's material identity, was functionally equivalent to all bodies, and thus robbed of its specificity. Those bodies that were unable to work, due to age, illness, or disability, were not in the Lockean sense individuals. Because they were held to dependence on others for the fundamentals of human life, these individuals had no agency. And, as Paul Youngquist has pointed out, "bodies irreducible to functional norms live beyond the pale of liberal politics, the objects perhaps of charity and affection but not quite persons, not quite proper. They remain too dependent to be full participants in civil society" (Youngquist 2003, 21).

In equating labor with the property thus accumulated, Locke goes further, reshaping the body that produces (and grounds) labor from something diverse and various (in its power and value) into something standardized and functionally equivalent inasmuch as all forms of laboring bodies produce property. "Not possessions but the ability to possess is what qualifies the individual for participation in civil society. It is because a man has a right to the free use of his body that he can accumulate property, which civil society then develops to protect" (Youngquist 2003, 20). Race, gender, ability, and even age are all implicitly specified in this understanding of property, which is focused not on possessions but on the power to possess. With that move, the citizen is redefined. Those social technologies that assisted in the transformation of the citizen would also help to transform farming over the next three hundred years from a way of life to a profit-driven business, or agribusiness.

Just as our understanding of the manifesto genre is deepened once we know that it originated in the struggle against agricultural dispossession and the redefinition of the citizen in terms of the capacity to possess property, so too we can deepen and animate our appreciation of other literary and aesthetic genres when we pay attention to agriculture as a set of practices, technologies, and actors (both human and animal). Yet despite Locke's explicit linkage between agriculture as the crucible in which a new citizen and subject was forged—so central to any cultural studies analysis—until recently it was left to such social science fields as agricultural economics, history, anthropology, geography, and rural sociology to challenge the metropolitan mindset of cultural studies (Cloke and Marsden 2006; Hart 1998; Levidow 1996; Sachs 1996; Soper 1996; and Thompson 2007). One such challenge originates in the program in agrarian studies at Yale University, an experimental, interdisciplinary academic program formed in 1991–1992 with support from the Rockefeller Foundation and the Ford Foundation as well as Yale University itself. The Yale agrarian studies program explicitly reaches beyond its social science origins, eschewing the "purely statistical and abstract," and setting an agenda for itself that welcomes "the fresh air of popular knowledge and reasoning about poverty, subsistence, cultivation, justice, art, laws, property, ritual life, cooperation, resource use, and state action."[7] Intending to draw together a wide range of disciplines, agrarian studies embraces three shared principles: "that any satisfactory analysis of agrarian development must begin with the lived experience, understandings, and values of its historical subjects"; "that the study of the Third World . . . must never be segregated from the historical study of the west, or the humanities from the social sciences"; and finally that "the only way to loosen the nearly hegemonic grip of the separate disciplines on how questions are framed and answered is to concentrate on themes of signal importance to several disciplines."

Agrarian studies is explicit in its intention to include the humanities in its social sciences–centered area of investigation, if only partly successful in enrolling humanities-based practitioners. Yet we need more than agrarian studies' attention to "rural life and society" broadly conceived, as I demonstrate in what follows. We need the explicit, critical, and cultural analysis of agriculture that can reside in an agricultural studies perspective. While agrarian studies draws its strength predominantly from the social sciences, despite its more far-reaching intentions, agricultural studies can also enroll scholars in several emergent (and often interconnected) research areas that reach beyond the social sciences: science studies, animal studies, ecocriticism and environmental studies, women and gender studies, and science fiction studies.

Literature and science scholars have begun to use specifically literary methods to assess the impact of agricultural innovations on the individual and society. Thus, science studies scholars are forging a critical literature that explores the role of scientized agriculture in the production of the human being as citizen (Bryson 2002; Franklin, Lury, and Stacey 2000; Haraway 2008; Levidow 1996; and Thurtle 2007). Animal studies scholars explore not only the ontological otherness of animals and their interconnection with human beings but also the broader issues raised by the farming of individual animals (chicken, cattle, sheep); agricultural interventions into animal breeding and human innovations in assisted reproduction; and the broader issue of the co-construction of veterinary and human medicine (Agamben 2002; Squier 2006; Broglio 2008; Clarke 1998; Franklin 2007; Mitman 1999; and Ritvo 1997). Ecocriticism and environmental studies research explores the social cost of an extractive approach to nature whereas a subcategory styling itself "green cultural studies" adds the category "nature . . . plants, animals, elements" to such factors as "ethnicity/color, gender, sexuality, economic class, and age" that are all influenced by the impact of texts and social practices (Hochman 2000, 187; and Pauly 2007). Women and gender studies explores the meaning of agriculture to women: as a site of loneliness; as a zone of racist eugenics; as a mode of economic and technological exploitation demanding feminist activism; and as a site where gender and domesticity can be reworked in relation to a contested modernity (Casey 2009; Jellison 1993; Sachs 1996; and Weinbaum 2004. Finally, science fiction studies scholars have been making explicit the connections between science fictions about the production of food and reproduction of people, and contemporary agriculture (Franklin 1982; LeGuin 1996; and Squier 1994).

The Labor and Pleasure of My Chickens

This chicken abecedarium builds on all of those earlier works and draws from half a decade of the daily physical labor and pleasure of keeping birds of my own. My flock has numbered sometimes as few as five hens but, at the time of this writing, it has swelled to thirty-four hens (various bantams and assorted heavy breeds) and three splendid bantam Seabright roosters. These chickens have arrived at my door in a variety of ways, and each bird brings not only its own origin story but also its own social, theoretical, and ethical issues as well. There are the birds I bought at the live bird auction in Belleville, where Mennonite and Amish farm families mingle with local chicken farmers and suburban or rural families just getting into chicken raising. Several of those hens turned out to be ill with a respiratory virus, so I had to learn how to isolate, medicate, and care for sick hens against the backdrop of the avian flu scare. (They did not have H5N1 but a simple upper respiratory infection, according to the diagnostic avian vet at the local agricultural extension service.) After that, I had to think about whether I wanted to continue to buy hens, or sell my hens, at a poultry auction whose regulations were lax enough to permit the sale of infected birds.

Other birds came from a large-scale chicken hatchery known for the wide range of breeds on offer as well as their policy of not trimming birds' beaks (unlike other hatcheries that sell to the industry and routinely trim beaks to prevent crowded birds pecking each other). These birds were shipped to us at one day old, arriving at our post office at six in the morning, to the delight of the postmistress. We raised them in an old wading pool in our garage, under warming lights, until they were old enough to join the others in the chicken pen, or in another case to be put out in the meadow in a movable cage called a chicken tractor. Finally, there were those weak little Cobb 700s that I picked up from my university poultry science department, birds otherwise scheduled for "culling" because they were physically unfit to serve as model birds for agricultural extension classes in poultry exhibiting and judging.[8]

My encounter with those Cobb 700s led to thoughts about the ethics and aesthetics of designing life, the hidden practices of industrial agriculture, and the role of the global poultry industry in shaping not only what we eat but how we farm. Most of all, that encounter aroused pity for those grotesquely disproportionate birds I finally found collapsed in the hen house, dead of heart failure. Yet that first chicken encounter also led me to a sense of interconnection—with the people I have come to know in PERC, the Avian Diagnostic Laboratory, and the Program in Poultry Science, who shared my admiration for the birds they raised while maintaining their mission of serving the poultry industry; with those I met in the backyard poultry circuits, where pastured poultry and free range birds were the goal; with the children I met who were enthusiastic poultry exhibitors, breeders, or chicken keepers; and with those friends and strangers who responded with enthusiasm when I said I was writing about chickens.

The Interconnectedness of All Things

As I continued my research, I began thinking of my book as an abecedarium, or ABC. Hoping to draw attention to the bird's fundamental presence in American life—for chickens were once present in nearly every backyard—and to explore the fertile potential this humble domestic animal holds for all kinds of intellectual inquiries and practical pursuits, I decided to present my research in the form of a primer, a text traditionally offered to young children as their introduction to the world of education and the search for knowledge.[9] I soon found that any letter of the alphabet that I chose as my point of entry to a subject (whether "Augury" or "Biology," "Epidemic" or "Inauguration") connected me to issues that reached beyond agriculture, issues I found myself exploring in the other chapters as well.

Interconnectedness, Zen Buddhism teaches, exists on all scales, from the very small and personal to the universal. This concept explains "the way suffering can arise in our lives, and the way it can end."[10] This may seem a rather high-flown formulation of my experiences while writing this book about chickens.[11] But consider this analogy used by Burmese Buddhist teacher the Venerable Mahasi Sayadow to describe the interconnectedness of all things:

> For example, the fate of chickens and ducks is terrible. Some are eaten while still in the eggs. Even if they hatch, they live for just a few weeks, and are killed as soon as they put on sufficient weight. They are born only to be killed for human consumption. If it is the fate of living beings to be repeatedly killed like this, then it is a very gloomy and frightful prospect. Nevertheless, chickens and ducks seem content with their lot in life. They apparently enjoy life—quacking, crowing,

eating, and fighting with one another. They may think they have plenty of time to enjoy life, when in fact they may live for just a few days or months.   (Sayadow 1999)

Interconnectedness is one of the most profound truths of Buddhism. Some practitioners refer to it as "dependent origination," but Vietnamese Zen Buddhist Thich Nhat Hanh calls it "interbeing" or "interdependent co-arising," and he defines it with apt simplicity: "According to the teaching of Interdependent Co-Arising, cause and effect co-arise (*samutpada*) and everything is a result of multiple causes and conditions. The egg is in the chicken, and the chicken is in the egg. Chicken and egg arise in mutual dependence. Neither is independent" (Hanh 1998, 221). As I followed alphabetical order to explore different facets of the chicken and the egg, several links on the chain of interbeing emerged as major themes: ignorance, formations, consciousness, feeling, becoming, birth, aging, and death. Without pretending to illuminate their specialized meaning in Buddhist thought but taking them rather as illuminating metaphors, I want to conclude by suggesting some of the meanings these aspects of interbeing acquired in this exploration of chicken culture.

Ignorance contributed in crucial ways to the experience of writing this book, not only the practical ignorance about chickens, poultry breeding, and agriculture with which I began the book but also the sense of disciplinary ignorance that greeted me with each new letter of the alphabet, each new topic, and each new chapter. I imagined this book as a primer about poultry, but I soon came to think of it as a primer in interdisciplinarity as well. To investigate the links between poultry science and chicken culture, and between science and culture more broadly, I had to resist the temptation to let my ignorance block my investigations. To work on this book, I took a vacation from the workaday academic culture of expertise, where knowledge is only available to those who already know. Conceiving of the book as a primer, I gave myself the holiday of curiosity. Late in the project, I rediscovered Eric Lott's defense of his own work process and celebrated a fellow spirit: "I have unquestionably poached on academic territory in which I can claim at best amateur competence. Writing this book has convinced me, however, that such an interdisciplinary attempt is worth the gamble and, especially given the habits of specialists and subspecialists, is an opportunity rather than an embarrassment" (Lott 1993, 9–10).[12] The pleasure in learning about the various cultural and scientific aspects of the relationship between human beings and chickens has been intense. Certainly, I owe this pleasure to my secure position as a tenured faculty member with joint appointments in several academic departments, but it also reflects an obligation growing from such multiple appointments: the commitment to remain open to the multiple impulses of curiosity fueling this project rather than retreating into the safety of one endorsed research area.

In its more specialized Buddhist sense, ignorance refers to the "misconception or illusion" that "makes us take what is false and illusory as what is true and real" (Sayadow 1999). I am tempted to apply this meaning to those moments of understandable ignorance bred of the distance we now live from the food we eat. For example, in response to a gift of a dozen eggs laid by my hens, a friend asks, "Can you actually eat them?" Another friend asks if I need a rooster if I want my hens to lay eggs. Acquaintances ask what a woman's studies scholar could have to say about chickens . . . or a science studies scholar . . . or a literature scholar. And a close colleague in New York City describes this book as "my Penn State book" which I have to "get out of my system" before I can get back to doing what he clearly thinks is real scholarship. The bunker mentality that distinguishes farm culture from scholarly culture, the country from the city, the body from the mind, and life from theory polarizes falsely but is easily remedied if we replace one view of ignorance with the other. As Shunryu Suzuki observes in *Zen Mind, Beginner's Mind*, "In the beginner's mind there are many possibilities, in the expert's there are few" (Suzuki 1970, 1).

Formations, to Buddhists, means "everything born or created." Clearly, the concept can refer to something as unitary and simple as Plato's featherless biped or as complex as the institutional structures for global chicken production in the twenty-first century. There is another way the aspect of formations was important to this book, however: the remarkably varied ways in which people responded to the chicken as a form. I explore the visceral meaning of formations to artists such as Dario D'Ambrosi and Miss Nancy Luce, who exhibit a phenomenological and embodied fellow-feeling for chickens: D'Ambrosi, in his play about Antonio, the disabled boy whose upbringing with chickens convinced him that he was a rooster, and Nancy Luce, in her empathetic understanding that it is as "distressing to dumb creatures to undergo sickness, and death, as it is for human, and as distressing to be crueled, and as distressing to suffer" (Luce 1875). In contrast, form[ation] seems an abstract concept for Jean Pagliuso, who claims to draw aesthetic inspiration from the negative space in a chicken's posture for her "Poultry Suite" photographs, rather than from the elegant fashion models or Hollywood movie stars the birds seem to resemble. Finally, [con]formation is both profitable and pleasurable to participants in the Chicken of Tomorrow Contest in the 1940s. The swell of a chicken's breast bespeaks not only consumer preference for white over dark meat but also the habitual, and meaning-laden, cultural association between women and chickens.

Formation can also mean a linkage between individual morphology and institutional structure, a convergence that can produce powerful social and political meanings, for good or ill, especially when deployed through the analogical structure linking chickens to human beings. We will see several different instances of this ideological deployment of form. Harry H. Laughlin's research program at the Eugenics Record Office in Cold Spring Harbor, New York, drew on experiments in poultry breeding to argue for negative and positive eugenic interventions in human reproduction, with horrific material results in the case of Carrie Buck. In contrast, Koen Vanmechelen's compendious theory of the "Cosmopolitan Chicken" aims to reshape the chicken biologically and artistically in order to stage his more metaphoric challenge to the Eurocentric and racist deployments of evolutionary biology and eugenics. Just as films of chick embryos are linked to the engineered breeding of whole poultry populations that would come later in the twentieth century, we will see that interest in morphology connects Camille Dareste's embryo experiments and Ray Bradbury's inspired chicken and her egg-borne message to Tod Browning's grotesque chicken woman and the embryology-obsessed restaurateur in Sherwood Anderson's "The Triumph of the Egg."

Consciousness, a capacity that is often used to differentiate human beings from the chickens they tend (and the featherless biped to which they were compared so long ago), plays a liminal role in these chapters: it is that which we have lost through our chicken farming practices, and it is the knowledge of the connectedness of all life that our encounters with chickens can enhance and extend. This book explores our loss of an awareness of the natural world gained by observing chickens, a skill evidenced in Elmer Boyd Smith's *Chicken World* (1910). His exquisite children's book maps the cycles of the human year, both agricultural and spiritual, on to the avian life cycle, ironically ending both the human and avian with the Christmas roast chicken. A later chapter explores the opposite perspective on nature—the risk consciousness that leads us to invest in scientific strategies for managing exotic threats while remaining unconscious of the very specific dangers we face daily. And yet these threats to mental and physical health produced by the economic restructuring of agriculture—the aggregation of poultry growing in large multinational corporations, leaving small growers to bear the risk while large producers harvested the profits—are far more likely to affect us than the risk posed by the avian flu.

Feeling appears, first of all, as "fellow-feeling": that powerful sense of embodied connection we have all but lost as the socioeconomic importance of that moral sentiment has been translated into the notion of rational preference so dear to supply-side economists. The onetime emotional and physical connection Nancy Luce felt to the chickens she raised was replaced by the adversarial response of Betty MacDonald to "hundreds and hundreds of yellowish-white yeeping smelly little nuisances which made my life a nightmare in the spring" (1945, 138), and the instrumentalism to which Linda Lord was reduced during her time on the killing floor at Penobscot Poultry. Of course this emotional trajectory is not unidirectional: the economic crash of 2008–2009 may have motivated some young people to involve themselves in Web-based volunteerism and community-supported agriculture, although fellow-feeling is still a scarce commodity in an economic downturn, as was indicated by the role of Massachusetts—a state with universal health coverage—in the health care bill debacle of January 2010.[13]

The notion of fellow-feeling can extend to race, gender, and ability, three categories that can bring us together or tear us apart. Beginning by granting that significance lay in a seemingly chance association between African American men and chickens, I found myself exploring the intersecting commitments to eugenics and racial privilege that have shaped the American legal tradition and set the challenging context for the presidency of Barack Obama. Other chapters also adopt the strategy of taking seriously the human–chicken comparison overlooked by Plato and challenged by Diogenes, exploring what it means to have one's humanity overlooked because of race, gender, or ability.

Becoming, or the process through which an embryo develops, was central to the tissue-culture research of developmental biologist C. H. Waddington. As he traced embryonic development and identified the organizer, he adapted film technology to reveal in time-lapse the changes resulting from embryonic grafting and incisions. And from the 1940s through the 1960s, the profit to be made by harnessing the process of avian becoming inspired U.S. grocers, poultry farmers, and filmmakers to collaborate on a competition to produce the "best meat-type chicken," a bird they claimed to be "ideal."

Birth provides the metaphoric focus for Stephen Lance's "Yolk." This film follows a young girl with Down syndrome as she come to grips with the meaning of birth and successfully negotiates the move from childhood to young womanhood, all by way of a school exercise in caring for a hen's egg as if it were her baby. Eggs can not only aid the production of an increasingly competitive and ideal normativity but they can also lead to the development of a nonnormate sociosexual becoming, whether for good or ill. Actual birth with a physical malformation becomes the occasion for Antonio's life-long isolation and his persistent delusion that he is a rooster, confirmed for him when his yellow plastic chicken lays an egg.

Aging, dying, and death. Human beings and chickens have been most closely united and most bitterly divided by these final three aspects of life. Whether we are Nancy Luce or the oldest rooster from the Cosmopolitan Chicken Project, kept by Koen Vanmechelen in a "hospice" chicken run, we age and are subject to the biological and social implications of that inevitable and often painful fact. We must die, every one of us. But here the unity between human and avian splinters into ethically charged and politically challenging differences.

Instead of attempting to smooth out these complex differences or to adjudicate them, I offer this book simply as a partial primer. Partial in the sense that its chapters cover just part of the alphabet, spanning A through I and then, after a gap, Z. And partial in that each chapter is distinctly and unashamedly *partial* to its subject. In fact, despite the warnings of E. B. White, in writing this book I risk my own share of withering glances to admit that I am very partial to chickens—and what we can learn with them, from them, and about them. While my own feelings about keeping chickens mostly take a back seat to the cultural, historical, and theoretical issues I explore in the chapters that follow, I hope that as I move between poultry science and chicken culture, my efforts reflect enough of the textures of thought and feeling to offer a satisfying response to the question: "why chickens?"

Augury

"There's no putting a human price on a thing like that."
–Ray Bradbury, "The Inspired Chicken Motel"

6:45 A.M. One chick jumps on the hen's back. The hen ruffs up her feathers, crouches down, fans her tail feathers, and spreads her wings. All ten chicks run under her tail and backfeathers, pressing in as tightly as possible. The two Cobb 700s freeze still and silent right where they are standing, on the little hill next to the chicken shed, and the Buff Orpington is quiet in the tall grass. I wonder what is happening. Then I see that a three-point buck, in velvet, has walked right up to the chicken yard fence. He raises his muzzle and sniffs the air in elaborate long snuffling sweeps, then tosses his head back and forth, moves in even closer, and does it again. I can see the rising sun reflected in his eyes. Then he turns and walks slowly away, down the fence line. The hen relaxes and stands, the chicks emerge, the Cobbs stir and walk down the hill peeping their strange high cries, and the Buff Orpington comes out of the high grass. Not even seven in the morning, and already I'm learning about the world from my chickens.

Augury, according to "bygone beliefs," is "the art of divination by observing the behavior of birds" (Redgrove 1920). The term comes from the Latin "augurs," the Roman officials charged with watching "the pecking behavior of sacred chickens" whenever a military or political initiative was under way, to see whether the gods approved.[1] For generations this practice of augury was understood as a supernatural phenomenon, as a moment of divine inspiration during a trance, or the result of expert work by a skillful reader of signs. But in 1920, when scientific explanations had trumped spiritual ones, augury was redefined by the British chemist and president of the Alchemical Society Herbert Stanley Redgrove as the supernatural *seeming* effect of natural phenomena: "Amongst the most remarkable of natural occurrences must be included many of the phenomena connected with the behavior of birds. Undoubtedly numerous species of birds are susceptible to atmospheric changes (of an electrical and barometric nature) too slight to be observed by man's unaided senses; thus only is to be explained the phenomenon of migration and also the many other peculiarities in the behavior of birds whereby approaching changes in the weather may be foretold" (Redgrove 1920, 34).

Natural, too, Redgrove believed, was the human impulse to find divine meaning in avian behavior, "to suppose that all sorts of coming events (other than those of an atmospheric nature) might be foretold by careful observation of [birds'] flight and song" (Redgrove 1920, 35). This attempt to substitute a scientific explanation of such moments of heightened awareness for the earlier supernatural one characterizes our era of reflexive modernization, in which new expert knowledge is always produced along with new modes of unawareness (Beck 1999). These new modes of uncertainty or ignorance are generated by the framework of our daily lives, in which we value technical, scientific, and rational approaches as the only ones we believe are capable of producing exhaustive and certain knowledge of ourselves and the world around us.[2]

Contrast to that scientific knowledge my morning experience in the chicken yard, or the experiences of those before me who, watching the scratching of chickens and other bird behavior, predicted the course of human affairs, foretold the future, or were simply surprised by a sudden insight or inspiration from beyond. Those generations relied on augury, a type of knowledge-making about the present and the future gained through intimacy with animals. This epistemological practice is in danger of disappearing in the twenty-first-century industrialized world. As we have accepted chickens as animals farmed for their meat and eggs in a process of rationalized scientific management, we have lost the ability to see what they and other animals augur for our collective future. We have deskilled ourselves of the subtle ability, whether we think of it as supernatural or natural, to find in animal behavior messages about ourselves and our world.

I mean "deskilled" in the sense used by Deborah Fitzgerald in her trenchant analysis of the rise of industrial agriculture, *Every Farm a Factory* (2003). Labor historians understand deskilling as the process by which a worker's ability, creativity, and autonomy are taken away by the introduction of sophisticated machinery into the workplace. Fitzgerald's contribution is to extend the notion of deskilling to a broader set of technical artifacts and procedures. She sees deskilling in the process by which new farming technologies have shifted the nature of contemporary farming from the creative management of land and livestock to the compliant application of pre-engineered products and practices. These new farming technologies are more than merely machines: they are networks of practices, assemblages of artifacts, and standardized operations. And these assemblages carry out activities once the purview of the human beings working alongside them. They have usurped the mental and physical agency of the farmer even while promising to extend it by enabling a higher yield for less money. To give some examples, such practices can include the creation of hybrid corn in lieu of a farmer's experienced selection of the right corn strain for the farm's specific soil conditions, the substitution of a milking machine for the skilled touch of a practiced milker, or the standardization of chicken raising in large indoor poultry houses rather than the farm wife's practice of raising chickens as one activity among many others on a diversified farm. Even as farmers learn the often welcome set of scientific, technologically based skills and practices that make up what Fitzgerald calls the industrial ideal of modern agriculture, we are also being deskilled in a set of idiosyncratically developed, often craft-based, and orally transmitted farming knowledges and practices. While family poultry raising still plays a large part in the local economy of the global south, where such traditional methods of poultry raising can still be found, until the recent growth of the backyard poultry movement they were fading rapidly from the industrially oriented poultry producing centers of the United States and Europe.

During the years that I have been raising chickens, I have made the transition back from supermarket culture to the chicken coop culture of my childhood and have realized just how much I don't know about chicken raising. To be sure, language retains much to guide the novice chicken farmer. So much lost knowledge is sedimented in our proverbs: that chickens do come home to roost, that my rooster does want to be king of the hill (or at least of the little mound of dirt and rocks in the center of the chicken yard), and most recently that I shouldn't count my chickens before they hatch. I received that poignant reminder when I went out to check on one newly hatched chick and found three others from the ample nest dead on the hen-house floor, one still half inside its broken egg.

Another kind of unlearning takes place when scholars who work in science studies and cultural studies ignore agriculture as a source of research questions. When we turn away from farming toward topics that seem more cutting-edge and exciting, such as genomics, biomedicine, cybernetics, or surveillance technologies, we surrender the knowledges, skills, choices, and modes of problem solving that were once a core part of the production of our food. When we assume that such issues are not significant subjects for our own scholarly research but rather can be delegated to the invisible infrastructure of our lives, we are being deskilled. We know so little about where our food comes from, and what little we do know we do not take seriously. This is particularly ironic since genomics, biomedicine, cybernetics, and surveillance technologies, as well as many other areas currently preferred over agriculture as subjects for science studies and cultural studies research, have been central to the massive transformation that has created industrial agriculture.

Two closely paired concepts from different disciplines can help to explain why this change has gone relatively unnoticed by cultural studies scholars: unawareness, a state of mind generated as inevitable byproduct to the creation of scientized expert knowledge, and agnotology, the inverse of epistemology; the deliberate production of ignorance. Both concepts draw attention to the negatively as well as positively productive effects of the disinformation generated by the search for scientific and technical knowledge and control. In sociologist Ulrich Beck's formulation, the production of unawareness enables the continuation of risk society with its built-in acceptance of a certain level of human costs to technoscientific rationality (Beck 1992). Agnotology, according to historian Robert Proctor, locates us in a state of willful not-knowing, alienating us from alternative epistemologies and knowledges that challenge the domain of scientific rationality (Proctor 2008).

The skill of augury can be recovered, however. Attuned by these notions of agnotology and the production of unawareness, cultural studies scholars have begun to return to agriculture, approaching it as a powerful source of analytic and activist strategies, and as one of the most significant sites of the global production of the human being (as consumer and producer, and as sexed and raced body, patient, citizen, and species).[3] Science studies scholars are returning to literature, a rich route into the social, political, and cultural imaginary. Literary critics are turning to stories, poems, and other works of imagination to discover that they preserve for reclamation modes of knowledge and significance once associated with farming while also articulating the tensions and fears aroused as innovations in science and technology have transformed that ancient human institution.[4] Three works of American literature spanning nearly a century can enable us to reclaim the power of augury as they explore the meaning of agriculture, the kinds of knowledge we have lost, and the modes of unawareness that have been produced as we have moved from "farm chickens to chicken farming" (Sawyer 1971, 26, 36, 112).

Arguably one of the foundational technologies of knowledge transfer, and certainly one of the developmentally earliest forms of social instruction, children's books convey essential knowledge from generation to generation. Let's begin, then, with a children's book that introduces us to chicken raising in the context of a whole ecosystem. Elmer Boyd Smith's 1910 *Chicken World* shows us life from the chicken's perspective. One April, "Old Black hen" decides to set (that is, to incubate) some eggs, now that spring has come, but her natural inclination is thwarted some by the farmer's intervention. "Though given a nest of eggs in a quiet spot she is suspicious and hesitates. 'I wish,' she grumbles, 'they'd let me do it in my own way. I hate a strange nest, and how do I know what kind of eggs they've given me. Those green ones don't look right.'" The coming of night, however, makes her try the nest anyway: "'No hen made this nest,' she scoffs. 'Still the eggs do feel very comforting, I daresay I can't do bet-

Now Spring has come, the Old Black hen decides to set. Though given a nest of eggs in a quiet spot she is suspicious and hesitates. "I wish," she grumbles, "they'd let me do it in my own way. I hate a strange nest, and how do I know what kind of eggs they've given me. Those green ones don't look right."

FIG. I. Old Black Hen Decides to Set.

Reproduced with the permission of Rare Books and Manuscripts,
Special Collections Library, the Pennsylvania State University Libraries.

ter'" (Smith 1910). Time passes, and the mysteriously colored eggs begin to hatch: "little chicks come struggling out. And long and last two little ducks. 'There,' says the hen in vexation, 'I've been cheated again, and ducks are such a care.'" The farmer has put some duck eggs in the nest, to profit from the hen's surplus labor. The tale follows the hatchlings, both native and alien, as they grow up, learn how to groom themselves, hunt for food by pecking around in the garden, establish their hierarchy in the barnyard by fighting with rival roosters, mate, and finally die.

Knowledge in this story is not merely the possession of the farmer, who built the nest with the mysteriously colored eggs for Old Black hen. Other varieties of knowledge, belonging to animals as well as human beings, are revealed as the narrative unfolds. We see the knowledge one species has of another, as when the Old Black hen realizes that the farmer has tricked her into hatching ducklings as well as chicks. There is species-specific knowledge, for example, when the Old Black hen relies on her knowledge of worms, bugs, and cats to teach her chicks and ducklings how to search for food and stay away from danger. As she is explaining that worms and bugs will "make little chickens fat and strong," suddenly a little kitten trots into their midst. The hen is alarmed: "'A cat!' cries the mother hen. 'Children never have anything to do with cats, they eat little chicks.'" And she draws on her own experience to frighten the kitten away, ruffling her feathers and flying at it "with a rush and a shriek" until it flees in terror. These knowledges and skills compete with each other right up the food chain, as one illustration makes clear. We see a man at work in his garden, tilling a patch of spring onions and another of cabbages. Unlike Old Black hen, who knows how to scare off an intruding kitten, the farmer does not know how to protect his territory. Although he pauses in his raking to wave, and possibly shout, at an intruding hen, his intervention is a failure. The hen keeps right on scratching for her food in his garden.

Finally, there is species-exclusive knowledge. "The ducks soon find the water tub, and in they flounder, to dive and swim, and will not listen to their mother who fears they'll drown. 'I hoped that this time I could teach them to be chickens,' she sighs, 'but it's no use, you can't change ducks alas.'" While in her wisdom the Old Black hen has access to knowledge philosophers would deny her, yet she still cannot persuade her chicks of its significance. Although fed by

But in vain, for they come to blows and fight a furious battle, and feathers fly about. Of course the young cock is blamed. "He really is very impudent," says the Old Black hen; "none of my children ever behaved so before." He is soon forced to run for his life, and then keeps out of the way for a while.

FIG. 2. They Come to Blows and Have a Furious Battle.

Reproduced with the permission of Rare Books and Manuscripts, Special Collections Library, the Pennsylvania State University Libraries.

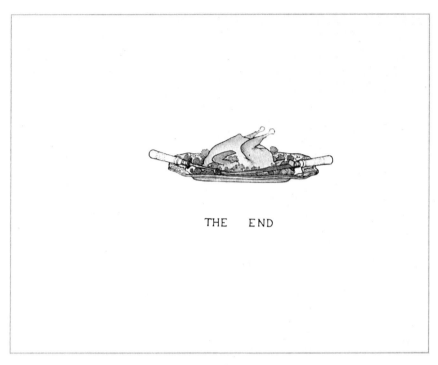

THE   END

FIG. 3. The End.

Reproduced with the permission of Rare Books and Manuscripts,
Special Collections Library, the Pennsylvania State University Libraries.

human hand, "the old hen still remains suspicious. 'I've heard that people eat chickens, too,' she warns, but of course they can't believe this and doubt their mother's wisdom."

In its wordless last image, which conveys a knowledge commonly held to be unique to human beings—the knowledge of death—the short story moves beyond the chicken world to consider the world all living beings share. Although children's stories nowadays would probably gloss over it, Elmer Boyd Smith's tale does not shirk from the harsh agricultural truth. The Old Black hen was correct in her suspicions: a roasted chicken graces the Christmas table. In all, this little turn-of-the-twentieth-century children's story conveys an expansive set of plural knowledges: of botanical and meteorological patterns; of diurnal, circadian, and seasonal rhythms governing animal behavior; and of ways the farmer can work with and disrupt the rhythms of animal life, producing ducklings as well as chicks by his obliging broody hen.

Chicken World seems to harmonize overlapping worlds: the avian with the botanical, and both with the human. The illustrations also incorporate the story's wider geographic and temporal context. Although our focus is on the chicken yard, we glimpse the world beyond: in the wild birds that stop by the chicken coop as well as in the chickens' two food bowls, one rustic local pottery, and the other Japan-ware. We follow the cycle of the blossoms in the fruit trees alongside the chicken coop and the succession of plantings in the garden. And we see the diversity of breeds once characteristic of chicken world, though rare in our day of identical white chickens. Old Black hen is a breed few modern readers would be familiar with—green legs, long skinny neck, rose comb, and wattles—far from the hybrid broiler cross of modern chicken production facilities. Yet the farmer's hand reaching into the nest shows us "Chicken World" on the threshold of change, its rhythms in the process of being disrupted. The tale augurs the coming of a new mode of chicken production, despite the hen's rueful comment: "I wish they'd let me do it in my own way."

The same year (1910) that Elmer Smith published *Chicken World*, a slim volume titled *Profitable Poultry Production* also appeared. This book offers an early glimpse of the rationalized, standardized chicken farming and egg industry of the future. Addressed "to the thinking farmers, farmers' wives and farmers' children," this guide to the "important practices and new wrinkles" of modern chicken raising asks the farmer for the first time to view the hen itself as part of the process of production, and to think about "not only the cost of the egg as a market commodity, but the cost of making the machine, the hen, which is to manufacture the egg" (Kains 1910, iv, 18). The shift in metaphor is also a narrowing: the hen is no longer a broody mother, a teacher, a scavenger for food, as well as a meal. Now she is simply a machine for manufacturing eggs. This model reframes the farmer's challenge as getting the most product out of the hen (eggs) for the least input into the hen (food, lodging, labor). We have come a long way from the notion that the hen might have her own perspective not only on the eggs she laid but also on her world.

While good farmers always tried to manage their livestock for productivity, the mechanistic model of chicken raising became firmly established after the U.S. farm crisis of the 1920s. When a drastic drop in the prices paid for agricultural crops occurred after World War I, coupled with a steady increase in the cost of other nonagricultural goods, farmers found their income failing to keep pace with their expenses. Many moved west in search of quick income. Yet as farmers moved to new land, they were challenged by new conditions in terms of transportation, irrigation, pest control, predator control, farming supply costs, and the availability of appropriate crop strains. Many farmers went bankrupt and lost their farms while others switched from sustainable farming to "extractive, go-for-broke farming" with damaging effect on the environment and on the farmers' well-being (Fitzgerald 2003, 18–19). As Secretary of Agriculture Henry C. Wallace put it, "In times such as these the problems of farm management on most farms are reduced to the simplest terms and can be stated very briefly. . . . Produce as much as you can and as cheaply as you can of what you can produce best; spend as little as you can; do without everything you can; work as hard as you can; make your wife and your children work as hard as they can. Having done this, take what comfort you can in the thought that if you succeed in doing what you set out to do . . . you will have produced larger crops than can be sold at a profit and you will still be under the harrow" (Wallace 1924, 2).

The farm crisis of the twenties was a challenge to the chicken industry. Farmers, bankers, civic leaders, and agricultural educators devised their own solution to the problem of 18 million unemployed. This was a business model for farming that stressed five elements imported from factory production: "large scale production, specialized machines, standardization of processes and products, reliance on managerial (rather than artisanal) expertise, and a continual evocation of 'efficiency' as a production mandate" (Fitzgerald 2003, 22–23). In the case of chicken farming, the result was the integration of poultry production (linking feed producers, and marketers) and contract growing (in which "the feed dealer furnished chicks and feed, and the grower furnished the house and labor") (Sawyer 1971, 85). Promises of a 50–50 split of the profits were tempting indeed to farmers who were hoping for "money in hens" (Boyer 1895).

That renewed drive to scientize and mechanize for increasing profits from poultry provides the background to the second literary text, Ray Bradbury's "The Inspired Chicken Motel" (1948). Unlike Elmer Boyd Smith's world of avian and agricultural diversity, Bradbury introduces us to a bleak new world of identical white chickens on a large-scale chicken farm. Set "deep down in the empty soul of the Depression of 1932," the short story tells of a poor family fleeing the farm crisis and the dustbowl toward the promise of California. Their stop for the night in a dumpy little motel outside of Amarillo, Texas, grants them a glimpse of the future that soothes their fears and nourishes their hopes. "It was, my father said, a motel straight out of Revelations. And the one strange chicken at that motel could no more help making said Revelations, writ on eggs, than a holy roller can help going wild with utterances of God, Time, and Eternity writhing along his limbs, seeking passages out of the mouth" (55–64).

The family decides to stop at the motel because of a sudden feeling of kinship with the chickens raised there, as fellow victims of the Dust Bowl. They see "an old man boot a rooster and smile as he came toward the auto gate" and they feel right at home "because we found that chickens are kicked the same as families kick each other, to get them out of the way" (57). Unable to sleep in the hot Texas wind, the family is visited by a Sybil—"the landlady, a frail woman whose picture I had seen in every newsphoto of Dust Bowl country, eroded down to the bones but with a fragile sort of candlelight hollowed in her eyes. [She] came to sit and chat with us about the eighteen million unemployed and what might happen next and where we were going and what would next year bring" (57). Bradbury's description of her face captures loss and desolation but also endurance and hope: "It was a face in which her life and the life of her husband and the ranch they lived on struggled to survive and somehow managed. God's breath threatened to blow out her wits, but somehow, with awe at her own survival, her soul stayed lit" (58).

After conversation has run its course, she leaves the room, reappearing with two small gray cardboard boxes: "The way she carried them it almost seemed she was bearing family heirlooms or the ashes of a beloved uncle." She opens the boxes, slowly and dramatically. In the first one, a small egg lies on a bed of cotton saved from an aspirin bottle:

> There, in the center of the egg, as if cracked, bumped and formed by mysterious nature, was the skull and horns of a longhorn steer.
>
> It was as fine and beautiful as if a jewelsmith had worked the egg some magic way to raise the calcium in obedient ridges to shape that skull and those prodigious horns. It was, therefore, an egg any boy would have proudly worn on a string about his neck or carried to school for friends to gasp over and appraise.   (59)

That longhorn emblazoned egg, recalling the days of open-range cattle ranching, seems out of place in the dusty outskirts of Amarillo, where the cowboy dream of freedom and self-reliance has dwindled to a sad attempt at chicken farming. But the second egg redeems the lost promise of the frontier as it offers them an augury of their future:

> There were words written on this egg in white calcium outline, as if the nervous system of the chicken, moved by strange night talks that only it could hear, had lettered the shell in painful half-neat inscriptions. And the words we saw upon the egg were these: REST IN PEACE. PROSPERITY IS NEAR. And suddenly it was very quiet. We had begun to ask questions about that first egg. Our mouths had jumped wide to ask: How could a chicken, in its small insides, make marks on shells? Was the hen's wristwatch machinery tampered with by outside influences? Had God used that small and simple beast as a Ouija board on which to spell out shapes, forms, remonstrances, unveilings? But now, with the second egg before us, our mouths stayed numbly shut.   (60)

The egg's augury is in stark contrast to its surroundings: no massive cattle ranch stretches out away from them, but instead the miniature Dust Bowl version, a *chicken ranch*, anchor for the hopes of so many farmers who had lost their land and livestock, and were trying to make one last go at farming. The landlady points out the chicken that laid that portentious egg: "We all looked over at ten thousand chickens veering this way and that in tides. . . . And for a brief moment I thought I *did* see one chicken among many, one grand bird whiter than the rest, plumper than the rest, happier than the rest, faster, more frolicsome and somehow strutting proud" (62). The inexplicable message draws together the bickering family, and as they drive on the following day, they speculate on its meaning. The father muses, "Ten thousand dumb chickens. And *one* of them, out of nowhere, takes it to mind to scribble us a note." But what is the message? When the son asks how much an egg like that is worth, the father's answer makes the augury plain: "'There's no putting a human price on a thing like that,' he said, not looking back, just driving for the horizon, just going on. 'Boy, you can't set a price on an egg like that, laid by an inspired chicken at the Inspired Chicken Motel. Years from now, that's what we'll call it. The Inspired Chicken Motel'" (63).

Bradbury's story demonstrates the distinctions through which Depression-era agriculture reinvented itself: between the individual and the population (that uniquely inspired chicken versus the ten thousand identical ones); between the notion of market price and the philosophical and spiritual notion of "worth" (the value of eggs on the commercial market versus the value of these inspired eggs, so rare that "there's no putting a human price on a thing like that"); and between the increasingly prevalent notion of farm animals as unthinking machines and the older sense of their mystery and wisdom: "Some creatures are given to talents inclined one way, some another. But chickens are the greatest dumb brute mystery of them all. Especially hens who think or intuit messages in calcium-scrawled froth in a nice neat hand upon the shells wherein their offspring twitch asleep" (55).

Bradbury's short story dramatizes the painful difference between the present-day reality of debt and unemployment and a future of economic promise that the egg predicts: "REST IN PEACE. PROSPERITY IS NEAR." The inspired chicken, at the Inspired Chicken Motel, speaks to the fantasy of the industrial ideal undergirding the transformation of agricultural practice in the 1920s and 1930s.

Scientific and spiritual auguries war with each other in the final short story I want to consider, Ruth Ozeki's "The Death of the Last White Male" (2006). From its title onward, Ozeki's story offers an explicit and focused example of augury, investigating a small rural event that holds powerful portents for the future of humans and animals. Ozeki's story takes place in the chicken yard, yet this time it is neither the site of domestic education nor of industrial inspiration but simply a site of labor, both productive and reproductive. On the first day of the Year of the Cock, the Chinese New Year, a migrating hawk kills the gorgeous white Silkie rooster that Chinese American–Canadian émigré Grace has been raising in her rural backyard. Her hens are both pets and producers: sex workers, she thinks, whose job is to produce eggs. Product yield—a quantifiable measure of economic success, seems to be the central anxiety of this story, but as we come to learn, Gimp's death augurs more than merely economic ill luck. As the story begins, however, Gimp's demise seems to Grace just the culmination of his longstanding role as an inadequate manager of an insufficiently rationalized production site. Wounded in the fight in which he vanquished the previously dominant rooster, "he took over running the white hens, but he was never quite up to the job, possibly due to his crippled leg. He could never quite get on *top* of things, Grace complained, and after he took over, the number of fertilized eggs dropped precipitously, as did the number of new white chicks. In fact, the survival of the entire white population had been looking iffy for some time now" (2006, 61).

Even a living, wounded Gimp made the prospects for the Silkie flock seem "iffy," but when he fails to protect his hens against a migrating *Buteo lagopu* hawk and ends up "a flocculent white pile, which she took to be a small drift of melting snow," his managerial capacity, like his potency, is at an end (63). His passing—the death of the last white male of the title—not only poses a threat to the population of white Silkie chickens but also resonates out to other threatened populations as well.

In "Death of the Last White Male," as in *Chicken World*, fiction is the vehicle for analysis of environmental, social, and economic changes as intersecting parts of an ecosystem. Ozeki's story also asks us to think about the relationship between modes of production and the health of populations. Risk discourse saturates the story. Grace and Bob have fled the risk of terrorism after 9/11, moving north from New York to Canada, and yet fear still hangs over them like a toxic fog. And there is the risk of pregnancy itself, all too evident to Grace during her gray February gestation, though she still hopes for a sunny May birth. The boom in real estate development has a harsh impact on this rural island near Vancouver, where both the wildlife population and the surrounding habitat are endangered. Grace's husband, Bob, is helping a researcher who studies one kind of die-off, the gradual extinction of a species of water shrew, *Sorex palustris brooksi*, while another kind of die-off happens all around him. Houses replace habitats, and human beings replace *insectivora*. And as the shrews respond to their narrowing niche, viral mutation expands and the H5N1 strain of avian flu proliferates. This frightening new strain can mutate as it moves from birds to pigs—and perhaps even to human beings—and as it travels from continent to continent, carried in the feces of migrating birds such as Grace's Alaska-bound *Buteo lagopu* hawk.

Hybrids proliferate, from the H5N1 virus to the Asian-bred Silkies Grace and Bob raise, and even to Grace and Bob's baby to be: all of them the products of nature–culture convergences. By engineered or spontaneous genetic modification, by air travel, and by globalized trade (what used to be called import–export) these scientific, geographic, economic, and sociological processes are "pivotal to the new universalisms of global culture" (Franklin, Lury, and Stacey 2000, 10, 73–74, 99). Yet they are a mixed blessing, as we soon learn. Global economic development, which drew Grace and Bob to British

Columbia in search of better health care, has reduced populations in East Asia to ill health, poverty, and marginalization. And although real estate development may bring jobs and houses to this island off Vancouver, it is harming the habitat of shrews and hawks as well as humans. Finally, the avian flu poses not only an economic risk for Grace's aunties and uncles in Hong Kong—the closing of live bird markets and the end of a livelihood—but also a health risk that could extend beyond Southeast Asia to the Pacific Northwest, and to Grace, Bob, and their baby to be. Grace recalls frightening television coverage of white-suited workers gassing and burying live chickens. "If the virus continued to mutate and develop the ability to mix or swap its avian genes with human genes, the flu could infect 25 to 30 percent of the world's population. Hundreds of millions of people could die" (66).

The millennial anxiety that shadows Ozeki's story fuels Grace's hunger for auguries. Unlike the calcium writing on the inspired egg, however, the auguries available to Grace throughout the story are ambiguous. The oranges Bob feeds her on New Year's Day—a cherished Chinese ritual they have optimistically relocated to the land of redwoods—disappoint, as out of place as Grace is herself. Trucked in at the wrong season of the year from a far-away orchard, they are tasteless exotics. And the oranges' out-of-placeness resonates throughout the story, to Grace's position as the daughter of immigrants who hasn't really gotten in touch with her Chinese roots; to Grace and Bob's status as recently exiled New Yorkers and as newly naturalized Canadians; to the Silkie chickens, far from their land of origin; and to *Buteo lagopu*, the rough-legged hawk, and H5N1, two versions of a death-dealing migrant. At the story's end, the endlessly optimistic Bob promises Grace that spring awaits: "Just you wait for them plums." But Bob's scientific knowledge jostles with Grace's uncertain intimations of the changes—climatological, sociological, personal—that are awaiting them. Despite his assurances, the future seems to hold more dread than savor as the Year of the Rooster begins.

Prophetic realizations take different forms in these stories about chickens, and so do the kinds of close attention that the chickens receive. There is the naturalist observation of *Chicken World*, which gives rise to accurate representations not only of chicken behavior but also of the flora and fauna around them on a small farm. There is the superstitious desire that pulls meaning from random chaos in Bradbury's story, finding the inspired chicken and her revelatory eggs among the ten thousand identical chickens on that Dust Bowl chicken ranch and hoping to acquire from them an economic and spiritual agency unavailable to this family struggling with Depression-fueled structural readjustment of agricultural income in the United States, from small private farms to large corporate farms. Finally, in Ozeki's story, there is a dramatization of how the forces of scientific and spiritual knowledge *and* unawareness link culture and nature in the management of a small chicken flock, or a shrew population, or a nation, or a human family. In each work of literature, close observation of bird behavior enables greater understanding of the human and animal habitats we share, providing auguries worthy of our attention.

Moving from matters of fiction to matters of fact, we can even find a transformed set of eggshell auguries operating in a rather surprising venue: the poultry industry itself. Public concern about food safety increased in Great Britain at the turn of the twenty-first century following the disastrous bovine spongiform encephalopathy (commonly known as "mad cow disease") outbreaks of the late 1980s and 1990s; a salmonella outbreak linked to imported poultry; and, most recently, intense media focus on the avian flu. In response to these events, in 1998 the British Egg Industry Council introduced its "Lion Quality mark," to be stamped on egg shells and egg cartons as an indication that the eggs were produced in Britain to specific quality standards that included "compulsory vaccination against Salmonella Enteritidis of all pullets destined for Lion egg-producing flocks, independent auditing, improved traceability of eggs and a 'best-before' date stamped on the shell and pack, which shows that they are fresher than required by law, as well as on-farm and packing station hygiene controls."[5]

The rampant Lion brand was joined by another logo when the British food industry established Assured Food Standards (AFS) to unify all of the food security claims or "food assurance labels." Representing itself as working "with the entire supply chain," AFS developed the Little Red Tractor as its new trademark, "a definitive, easy-to-spot symbol of assurance that food had been farmed and produced to independently inspected standards."[6] The AFS Little Red Tractor trademark was "launched by the Prime Minister on 13 June 2000," and modernized five years later when the word "Little" was dropped "to give a stronger image," the word "Farm" was replaced on the Web site by the word "Food," and the word "British" replaced by an image of the Union Jack. This revamped Red Tractor Logo is now imprinted on all eggs produced in Great Britain.[7] Established as a guarantee of food safety (constructed as linked to the notion of a nationally branded product) the Red Tractor Logo is surrounded by the words: "Assured Food Standards GREAT BRITISH CHICKEN."[8] This food safety guarantee relies on the sciences of statistics and probability theory, anchored in the linked practices of risk analysis, insurance (or assurance, as it is called in the United Kingdom), regulation, and standardization. Responding to the anxiety-inducing phenomena of globalization (including but not limited to the fears of avian flu also registered in Ozeki's story), the Red Tractor Logo returns the consumer from a world of global egg trade to a reassuring (if fantasized) scientifically monitored nutritional and economic nationalism.

The Red Tractor and Lion Quality Mark egg stamps supply scientific assurances of safety rather than the mystical promise of well-being provided by Bradbury's inspired hen. The change in food production acknowledged in the revamped Red Tractor Logo—the turn from the specificity of "farm" to the delo-calized, generic "food"—blunts our abilities to appreciate and respond to the complex range of meanings embodied in chicken farming. Such deskilling is taken to the extreme in the British Egg Industry's announcement on July 31, 2006, that "they've cracked it!—the self timing egg." Promising to do away with the "age old problem of cooking your boiled egg just right . . . step aside the clock and the running sand granules" the BEI Web site explains: "Using the lat-est head sensitive technology, Lion Quality egg producers tasked boffins [scien-tific experts] to create a foolproof way to guarantee the yolk comes out the way you want it—soft, medium or hard boiled—every time. Carrying a special label, the egg goes into your pan as normal and when it is perfectly cooked the Lion logo appears as if by magic confirming your chosen preference."

From magical etched eggs to British Poultry Board egg stamps that promise both science and magic, we have come to the end-point (for now) of a trajec-tory away from the notion of augury I have been exploring in this chapter: a mode of learning—whether supernatural, spiritual, or natural—grounded in human observations of animals. Redgrove the alchemical chemist claimed augury as a natural response to animal behaviors so subtle as to escape con-scious notice. He argued that auguries are not messages from the divine but rather messages from ourselves: the selves that still exist in possibly obscured relation to our animal companions on Earth, even in this rationalized, scien-tized era.

What is augury, now? Augury survives as a metaphor for the modes of awareness, the knowledge-making practices that are being lost to human beings as we move into what is increasingly an instrumental relation to ani-mals. Sitting with my chickens, I have come to appreciate the things I can learn from that simple awareness: the presence of a hawk above a flock of chickens suddenly still, or the threat posed by a backyard deer to new hatchlings. While we are losing this knowledge every day as our food sources move farther from us, we can still access it in works of fiction such as these three short stories: they reveal the crucial importance of agriculture to human health, ongoing life, and cultural ritual. Such fictions teach us what these poultry and egg industry Web sites obscure: that a hunger that is more than simply material connects the human who feeds the chickens to the chickens that feed the humans.

Biology

"Of all the strange things in biology surely the most striking of all is the transmutation inside the developing egg, when in three weeks the white and the yolk give place to the animal with its tissues and organs, its batteries of enzymes and its delicately regulated endocrine system. This coming-to-be can hardly have failed to lead . . . to thoughts of a metaphysical character."

–Joseph Needham *A History of Embryology*

"Go to work on an egg" urged the British Egg Board in 1990, in advertisements showing dark-suited men and women riding giant hen's eggs down crowded city streets. I was living in London that year, reading *Guardian* reports on Parliamentary debates over human embryo research, and the Egg Board ads made me laugh. What an unfortunate choice of slogan! In the years since, however, that slogan has seemed prophetic, challenging the narrow focus on human eggs and embryos with a broader vision of reproductive control as central not only to human medicine but to agriculture as well. From biologists and biomedical researchers to cinematographers, industrial designers, and supermarket buyers, people in a range of professions have indeed gone to work on an egg, drawn by its remarkable features as an experimental model: the easy access it provides to the malleability and plasticity of life itself.

In November of 2007 I returned to England to discuss egg engineering with people intimately involved in the practice, though in two very different contexts. I spent a day in Oxfordshire with the group of industrial designers who were inspired to create the first high-design urban hen house, the "eglu," and the following day I met with Claudio Stern, J Z Young Professor and chair of the Centre for Stem Cells and Regenerative Biology at University College, London. Stern had intrigued me by claiming, in a 2005 article, that the chicken, "one of the most versatile experimental systems available," had now "come of age as a major model system for biology, medicine, and agriculture" (Stern 2005, 16, 1). My conversations with the graphic designers and the biologist sparked not only metaphysical speculations but also some very material appreciation of the complexities of chicken culture, those processes by which the chick embryo, cinema technology, and chicken farming were harnessed to produce "the Chicken-of-Tomorrow."

Move Over, iPod: Here Come the Chickens

Our rental car jolted along the pitted road around the back of a suburban house to a small industrial park, where we came to a stop next to a Prius. This was the Oxfordshire headquarters of Omlet, a small British corporation that manufactures the hen house I would vote "most likely to make it into the Museum of Modern Art." As we pulled in, a small cardboard box with large air holes from which clucks and squawks emanated was being loaded into the open hatchback. A chicken was about to be delivered to a new eglu owner, we learned from Johannes Paul, the lanky, bearded young man who met us at the door. After having earned an undergraduate engineering degree, Paul wound up at the Royal College of Art in London, studying industrial design, together with his friends James Tuthill, Simon Nichols, and William Wyndham. As Paul explained, "In our final year we were looking at working together afterwards and starting a company, and [trying to identify an area which] would distinguish us from people who were just doing furniture, tables and chairs. . . . And James kind of pops up with this idea of Why don't we redesign a chicken house?"[1] Johannes was still animated by the excitement of the moment he described: "[We] could see how we could develop not just the product but a whole kind of community around it and we could build this lifestyle and give people something they genuinely wanted and needed, and we could repackage chickens in a kind of twenty-first century way, which would mean that they were accessible again for the vast majority of people. . . . Now they would go: I really want to keep chickens, and I've found this great company who make this amazing chicken house and you can buy it all online and have it delivered right to your garden. *With the chickens*" (Paul and Wyndham 2007).

Their new product was the eglu, a glossy plastic chicken cage in the shape of an egg with an attached screened run that allowed the chicken access to the green grass of a backyard. Arguably the biggest development of the new century in urban chicken raising, the eglu was launched at the Royal College of Art Show at London's Olympia in 2004, where it received huge press interest. As Paul recalls, the headline in the London *Times* read "Move Over I-pod: Here Come the Chickens." The new media allusion is apt, because the eglu is an Internet-based product, along with the Eglu cube (with or without run and extension) and a full range of feed, domestic items, books, gifts, accessories, and clothing that are also available online at the Omlet Shop.[2] Now distributed in the United States by the Murray McMurray hatchery as well as directly by Omlet in the United Kingdom and Europe, the eglu is a completely packaged concept for chicken raising. The buyer chooses the color of eglu she prefers (they come in red, orange, blue, green, pink, and purple; a less eye-pleasing version in brown was discontinued due to low sales) from a glossy, beautifully designed brochure or Web page, and she chooses her new chickens as well, which are delivered directly to her door by the Omlet company along with her new hen house: "We deliver and install the eglu complete with organically fed chickens throughout a large part of the UK. . . . If you live within the chicken delivery area you can choose from 2 delightful breeds of chicken. The magnificent Gingernut Ranger and the chick Miss Pepperpot are both excellent egg layers and friendly pets." Moreover, for people who need support in this new farming endeavor, the Omlet team is there, at the Omlet Web site, with its Omlet club, "where people with lots of experience of keeping chickens in eglus share tips and advice with people eagerly expecting their first egg."[3]

Evidently, Johannes Paul and his colleagues have accomplished a major act of aesthetic and biological egg engineering. From chicken coop to Internet community, they have re-created chicken raising as a high-concept, high-design experience rather than the culturally alternative, back-to-the-land lifestyle one might have expected. Incorporating the user-friendly model and rhetorical tone more characteristic of pregnancy and parenting Web sites than the farm-and-feed journals, the Omlet Corporation has truly gone to work on the egg, repurposing it beyond simple reproduction, or even basic food production, to entertainment as well. As Paul explained over a local foods lunch at a nearby pub where the influence of Hugh Fearnley-Whittingstall and River Cottage were in ample evidence, "We don't sell people this idea that you keep chickens and you'll have this amazingly healthy lifestyle, we sell people an entertainment package. It's a bit like Sky Movie Channel; this is the Eglu Movie Channel, and you can go out and you can watch your chickens, and the eglu's really cool, and it's kind of a leisure product as well as being a food" (Paul and Wyndham 2007).

Hens as entertainment, the hen house as TV screen, farming as the production not merely of food but of leisure: I found plenty of material for metaphysical speculation as we drove back to London on the packed M40. Hens, eggs, and videotape. Chicken and egg inverted in Omlet's new order of life. Re-created in a punning rewind of the Humpty Dumpty story, the once-broken egg is now intact, widely available to customers around the world, and even larger than the chicken. The age-old question, "Which comes first, the chicken or the egg?" has its own branded sublunary corollary: "Which is delivered first, the eglu or Miss Pepperpot?"

"He Was My Hero . . ."

University College London is a less pastoral setting than the Oxfordshire countryside, and it was raining hard when I climbed the flight of stairs to the office of Dr. Claudio Stern, where—even indoors—the air still held a damp chill. However, as we sat amid stacks of biological journals, I forgot the cold as Dr. Stern and I discussed our shared interest in the work of C. H. Waddington. This polymath biologist carried out experiments on embryonic development that helped to make the chicken a valuable new model organism, part of biology's turn from simple observation to active interventions designed to provoke change and development. I had been friends for some time with Waddington's daughter, a brilliant mathematician and academic colleague, who had shared with me stories of his love for modern art and his metaphysical absorption in the meaning of what he called the "world egg." I had read about Waddington's remarkably wide-ranging research interests, spanning paleontology, genetics, developmental biology, growth, biochemistry, theology, and philosophy. More than a decade ago, doing research in the archives of a medical history library for my book on the literary backgrounds to the development of reproductive technology, I had delightedly discovered an example of that imaginative reach: a poem in which Waddington celebrates the pluripotent happiness of the newly fertilized egg: "Happy the egg, the bubble blastula,/ Does nothing much, but all the same/ Each part as good as every other" (quoted in Squier 2004, 78–79).[4] But now I was finding that Waddington was also central to the research of this major contributor to the science of chicken culture.

Claudio Stern was an admirer of Waddington, speaking of him as "my hero." Once, when he was an undergraduate, Stern had attended a meeting of the Royal Society and had even seen Waddington at a distance. In fact, he said, he had a "funny story" to tell me about Waddington. When in graduate school, Stern had been on his way to the train station for the trip to Edinburgh where he planned to begin study with Waddington when he ran into a friend: "He saw me with my luggage, and he said, 'Where you off to?' And I said, 'I'm just going to Edinburgh to see Waddington.' And he said, 'Haven't you heard? He died last Friday'" (Stern 2007). The sting of this missed connection with his hero may have been soothed somewhat when Stern was awarded the Waddington Medal for outstanding research and service by the British Society for Developmental Biology, but clearly it still lingered.

Conversation with Claudio Stern gave me a fuller sense of the crucial role Waddington played in the development of cellular and developmental biology. Stern praised his "unique ability to think very broadly about central problems and identify major questions . . . what seems like effortlessly" (Stern 2007). One of a handful of scientists who helped shift the focus of scientific experimentation from a logic that was substitutive to one that was "provocative, directed towards testing the limits of an interpretation of nature and living matter," Waddington's most important work concerned the mechanism of embryological development (Landecker, 2007, 57; and Stern 2007). Waddington spent the ten years between 1930 and 1940 working on the chick embryo, publishing numerous articles on the subject in the *Journal of Experimental Biology* and else-where, as well as his later landmark book *The Epigenetics of Birds*. On the title page of that volume, Waddington claimed as his own the new word he had coined, thus mapping a new terrain: " 'EPIGENETICS.' The science concerned with the causal analysis of development" (Waddington 1952).[5]

Waddington's research had an intuitive reach characteristic of the new era of biology, which now focused on exploiting the plasticity of living materials and "operationalizing biological time" (Landecker 2007, 11). Although he had published only "two extremely dull papers on the ammonite," after conversa-tions with Honor Fell, of the Strangeways Laboratory, and R. G. Canti, the early innovator of time-lapse photography of embryonic development, Waddington became interested in the chick embryo. Inspired by their new research approaches, in particular the use of time-lapse photography to record embry-onic development and the techniques of tissue culture, Waddington began to test different interventions in the developing embryo, working methodically and alone on a small number of simultaneous experiments. The resulting sequence of studies became classics in the field. "On the basis of that one embryo for each experiment, one case of each experiment, he published this one page paper in *Nature*," Stern told me with a laugh. "Now you can't do this anymore" (Stern 2007).

Waddington's genius grew on a disciplinary foundation, of course. Claudio Stern has assembled a table of "Some Major Concepts due to Work on Chick Embryos" that reaches from William Harvey's work in 1628 on the function of arteries and veins that led him to propose "the existence of capillaries"; Mar-cello Malphigi's study of the neural tube in the 1670s; and Wilhelm His's work on the neural crest in 1868 (Stern 2005, 11). As far back as 1895, Wilhelm Roux had accomplished the first successful "explant," or culture, of living tissues when he was able to isolate and grow an extract of chick embryo in saline solu-tion (Najafi 1983, 1086). And, as Philip Pauly has explored in rich detail, biolo-gist Jacques Loeb laid the groundwork of a new biology with his commitment to the "engineering ideal" for the life sciences (Pauly 1987). A sequence of experi-ments by Alexis Carrel and Montrose T. Burrows of the Rockefeller Institute in New York, and Ross G. Harrison of Yale University, advanced tissue culture work in the early years of the twentieth century, keeping portions of the embryonic chick alive in tissue culture for hours and then days (Carrel and Burrow 1910a, 1910b; Burrows 1910; Rous, Murphy, and Tytler 1912; and Rous and Murphy 1912; see also Squier 2004; and Landecker 2007, 68–106). Peyton Rous extended the application of tissue culture decisively into the realm of cancer research, reveal-ing in 1911 that cancers could be caused by viruses (Stern 2005, 11). Rous suc-ceeded in keeping a particular kind of chicken cancer—a fowl sarcoma—alive in culture for more than a year. Carrel and Burrows built on that success when they received, "through the kindness of Dr. Rous, two chickens with actively growing tumors," which they then excised and grew in culture. In 1910 they announced their achievement in "cultivating a very malignant sarcoma outside the body" in an experiment designed "to observe living cancer cells at every instant of their growth" (Carrel and Burrows 1910b, 1554).

Cancer cells were enormously important to researchers throughout the first four decades of the twentieth century; they studied these liminal life forms as key to the processes of normal and abnormal growth (Squier 2004). But in terms of public renown, easily the most famous and controversial of the experiments foundational to Waddington's later tissue culture research was announced by Alexis Carrel himself in autumn 1912. Newly the recipient of the Nobel Prize, Carrel told an astonished popular press that he had kept cells from the heart of a chick embryo alive and beating in culture for several months (Landecker 2007, 95; Najafi 1983; and Squier 2004). I have written elsewhere about the shocking transformation in the meaning of life and death that this experiment seemed to produce (Squier 2004). But now, as we consider the pivotal role played by the chicken as an object of biological and cultural engineering, the intriguing aspect is the nearly seamless way the press and general public applied to human medicine this rather esoteric experiment on embryonic chick cells. Beneath the headline, "Carrel's New Miracle Points Way to Avert Old Age," an unattributed article in the *New York Times* described his experiment as if it had been carried out on a human being: "This scientist has not only found a way to make connective tissue—the tissue of which the greater part of the body is composed—live permanently and grow outside the organism from which it was taken . . . he can actually regulate that growth."[6] Only near the end of the seven-column story did the writer acknowledge the extrapolation that had taken place, quoting Carrel's own prediction, "If human connective tissue cells could be preserved in a condition of permanent life, as connective tissue cells of the chicken are preserved, the value of the plasma of an individual might be appreciated by the cultivation in it of a group of these cells and by observation of the rate of their multiplication."[7]

This interspecies preoccupation with rates of growth, aging, and life immortal found their way into Waddington's experiments not only through the legacy of tissue culture in the first three decades of the twentieth century but also via some remarkable contemporary innovations in cinematography (Landecker 2007, 11). In 1913 the first film to be made of cells in tissue culture was produced by Jean Comandon, who had been a medical researcher until he formed his own film company with support from the Pathé film production company. "Survival of Fragment of the Heart and Spleen of the Chicken Embryo: Cell Division," was made in collaboration with two other biologists using the tissue culture techniques of Ross Harrison and Alexis Carrel (Landecker 2007, 85). In 1929 Ludwig Graper and Robert Wetzel made a film that was "one of the landmarks of the era" (Stern 2005, 9; and Bellairs 1953, 123–124). That film, now sadly lost, offered "a series of stunning stereoscopic time-lapse films revealing the movements of labeled cells in living, intact embryos . . . which also included observations of the behavior of isolated embryo fragments" (Stern 2005, 9). Photographer R. G. Canti, who after years in Alexis Carrel's tissue culture laboratory in New York City had developed a technique for filming living tissues through a microscope, was soon collaborating with Waddington (Squier 2004). With Canti and his colleagues Honor Fell and Gregory Pincus, Waddington carried out and filmed experiments on embryos in vitro, capturing in time lapse the minute changes resulting from excisions and grafts. Alexis Carrel praised film technology as essential to the new biological techniques of intervention in development, arguing in *Science* in 1931 that the old cell biology with its static method of staining and observing dead cells had been replaced by the new dynamic study of cells in time, because "time is really the fourth dimension of living organisms. It enters as a part into the constitution of a tissue. Cell colonies, or organs, are events which progressively unfold themselves. They must be studied like history" (Landecker 2007, 87).

Film technology was also important to science beyond the laboratory, where it was valuable in garnering support for the work of embryologists such as Waddington, Fell, and Pincus. They screened the films Canti had helped them make in movie theaters, scientific conferences, and even at 10 Downing Street, offering astounded viewers close-up looks at such experiments as "Cultivation of the Whole Chick Embryo Femur *in Vitro*" (Canti 1932). Transforming filmed records of their experiments into public film performances, Waddington and · his colleagues expanded the community of interest in embryo research, arguably enhancing the development of embryology as a scientific field.

The embryonic grafting techniques pioneered by Waddington in the 1930s can still be seen at work today, although the interpretations of the findings differ and even conflict with each other, reflecting the clashing models of contemporary evolutionary and developmental biology. We can sample the charged nature of this scientific debate in an article that appeared in the *International Journal of Developmental Biology* in 2009. The authors, a geneticist and epidemiologist who contest the current dominance of a "survival of the fittest" model of evolution, emphasize the importance of embryology not just "as a source of knowledge that illuminated our understanding of evolution" but also as a window into the processes of individual development. They argue that when seen from an embryological rather than evolutionary perspective, biological development more fundamentally involves collaboration rather than competition. Recent discoveries concerning the way embryonic cell signaling determines gene expression support the notion of a "cooperative genome" with growth guided by dynamically interacting genes and proteins rather than the canonical "selfish gene," which is seen as outcompeting all others in its struggle to reproduce itself.

This means that the genetic mechanism leading to cell transformation speculated on decades ago by Waddington and his colleagues is now being confirmed. This is the process of intracellular communication or gene signaling, through which "combinations of expressed gene products, and their arrangement in space and time, determine the state of a cell on the developmental scale of life." Through this process, Weiss and Buchanan explain, "cells are induced by signals to alter gene expression patterns" (Weiss and Buchanan 2009, 760). Microscopic though these findings are, their implications are huge because they suggest that the growth and development of an individual is shaped not only by signals in the genome but also by signals sent from cell to cell. Ironically, the authors observe, "the influence of environmental signals was much more readily accepted in the first part of the twentieth century than later on, when genes and determinism became predominant" (Buchanan 2009a).[8] What Waddington first glimpsed, and what further work would reveal, was that these chemical signals can be triggered in response to the environment, both inside the cell and outside the organism. It is these signals that determine which genes are expressed in response to a specific environmental influence (that is, for example, whether a BRCA 1 gene goes on to initiate cancerous changes). The study of cellular signaling and the resulting induced cellular changes carried out by Waddington and continued today not only gives us a close look at the unfolding development of life but it also challenges our simplistic assumption that genes alone determine behavior. As Buchanan put it later in a blog essay, "Darwin's . . . deep insight was that life is a history of process" (Buchanan 2009b). Examining the same process of induction and reception that Waddington studied in the early thirties, Weiss and Buchanan approach its significance on a scale that was unavailable to him: that of population genetics. As they explain, by "changing the scale of observation, from specific genes to the perhaps more amorphous nature of process" and reconceptualizing developmental biology through genetics, they have come to understand development as contingent, relational, adaptive and tolerant of variation (Weiss and Buchanan 2009, 760).

Here, as so often in scientific practice, words do not merely describe but help to bring into being. And here again, chickens prove good to think with. We know that metaphors were important to Waddington, who, when suggesting that "development is to be looked upon as the resultant of several forces," then added "the metaphor being applied as strictly as possible" (Waddington 1952, 227). Weiss and Buchanan also turn to metaphors to describe the process of signaling that organizes embryonic growth and the development of an organism: "All of these phenomena are *contingent*, in that the stage must be properly set in the proper order for the drama of sequestered, hierarchical cell-fate lineages to be established in embryogenesis" (Weiss and Buchanan 2009, 761). And as if testifying to the metaphysical implications of this particular exploration of biological performance, they break into metaphor as they stipulate that the signaling or coding elements in the genes "cannot be independent, but must instead be highly coordinated in space and time: there is no aboriginal egg followed by a subsequent chicken" (760). Just as attention to the environment around the cell helped Waddington to understand the multiple forces shaping the developing chick embryo, so Buchanan and Weiss's focus on cell signaling reveals a cooperative, rather than competitive, genome.

Surprising resonances linked my discussion with Claudio Stern to my conversation with the eglu designers, even though the focus had shifted from the eglu as a metaphoric TV channel to the avian egg and embryo as actual, riveting subjects for cinematographers. Central to both conversations were the experience of close observation and manipulation of living things and the technology of film. Capable of reordering time and space, bringing the past into the present, the present into the future, and the very distant object right under our noses, film was a central tool in the reengineering of life. Cinema technologies "helped materialize biological time as a thing that could be physically intervened in as part of experimentation with tissue culture" (Landecker 2007, 85). In biology of the thirties, cinema photomicography, or filming through a microscope, enabled researchers to stop time, speed it up, and run it backward to analyze the changes researchers induced in the developing embryo (Squier 2004). And in the early twenty-first century, the experience of watching hens in an eglu reminded graphic designer Johannes Paul of a television channel: its endless engrossing, captivating, and ever-new offerings for the eye as well as the eye of the mind. In the century between these two kinds of egg engineering, both using visualization technologies though in very different ways, World War II intervened, its ubiquitous newsreel footage bringing frontline experience to the home front. There, as in the postwar heyday of Hollywood cinema, film both catalyzed and recorded significant changes in chicken farming that accompanied the war effort and led to the rise of the American poultry industry.

Armies Don't Eat Chicken

On July 21, 1943, heavily laden trucks carrying crates of live chickens along Route 13 in Delaware from producers in Delaware, Maryland, and Virginia to markets in Washington, DC, were stopped by barricades just outside of Dover. Acting under the Second War Powers Act, and with the cooperation of the Delaware State Police, the U.S. Army had requisitioned their squawking cargo, "to provide needed food for soldiers" (Egan 1943). This unique use of the wartime power of the Army to requisition not rare metals but live chickens as "critical materials needed for war production" suggests how the advent of World War II changed the American view of chickens. In 1939 Frank L. Platt contributed an article to the *American Poultry Journal* asking "How Will the War Affect the U.S. Poultry

Industry?" (Platt 1939, 578). Platt concluded that "chickens are not a war commodity; their carcasses are too bulky to ship overseas, compared to a dressed hog. Armies don't eat chicken. The use of poultry and eggs, however, expands at home" (Striffler 2005, 43).

A "Food for Freedom" program pushed the new view that pork and beef should be saved for the soldiers. The only meat not subject to rationing, chicken, was left for civilians. This new Food for Freedom program capped a reorganization of poultry farming that began at the turn of the twentieth century when the Cooperative Extension Service, under the auspices of the Hatch Act (1887), sent poultry educators and extension agents out into rural communities to provide advice and classes in animal agriculture, vegetable growing, and home economics. The 4-H Club, founded in 1902 by Ohio educator A. B. Graham with the assistance of Ohio State University and its Agricultural Experiment Station, carried on a parallel educational mission for the younger population. Livestock raising, vegetable farming, canning, cooking, sewing, and housekeeping were taught in gender- (self-) segregated classes to rural boys and girls (Van Horn, Flanagan, and Thomson 1998). The 4-H Club focused on producing food during World War II, planting "victory gardens," and raising animals for meat. One scholar estimates that "from 1943 until the end of the war, 4-H club members produced enough food to feed a million men serving in the American forces." The 4-H Club members were seen by extension educators as "mediaries between the university researcher/educator and the farmer in the community. . . . Through the young peoples' involvement and accomplishments . . . the parents were exposed to new farming methods and were convinced to try and adopt new practices" (Van Horn, Flanagan, and Thomson 1998, 2).

Because the federal government had set chicken prices higher than the cost of production in a move to encourage the expansion of chicken farming, a black market in illegal chicken sales had sprung up. It was this illicit market that was the target of the army raids in July 1943: officials said it "had diverted food from the Army and deprived civilians in New York and elsewhere of poultry at legal prices" (Egan 1943, 13). By December 2, 1944, War Food Administrator Marvin Jones had announced an order to take effect on December 11 that would outlaw the sale to civilians of any chickens "produced and processed in the Delaware-Maryland-Virginia area," commonly known as Delmarva, to civilians "until the Army has obtained 110,000,000 pounds for fighting men and casualties in hospitals and recuperation camps."[9]

Eggs, however, were a different matter. The Army made significant advances in egg engineering in wartime. As Dr. Cliff D. Carpenter, president of the Institute of American Poultry Industries, recalled in his 1962 address to the Sydney, Australia, Rotary Club, "Egg drying became a dire necessity during World War II and made it possible to ship dried whole eggs all over the world—to help feed our Allies and our Armed forces. Today, now known as egg solids, they have entered world commerce and are shipped all over the free world" (Carpenter 1962). Posters linked the production of infertile eggs (which were thought to have a longer shelf life than fertile eggs) to the battle against fascism with slogans such as, "Two Jobs to Do: Beat the Axis and Swat the Rooster."[10]

Despite Platt's assurance that "Armies don't eat chicken," by the 1940s chicken had become part of the war effort. The Food for Defense program supplied chickens to the soldiers at Craig Field Southeastern Air Training Center in Selma, Alabama. The chickens were purchased from local Alabama farms where African American dressers, pluckers, and handlers played a crucial part in their production. Photographic records of these Craig Field chicken feasts, assembled by John Collier Jr., for the Farm Security Administration and the Office of War Information, document the white cadets who enjoyed the fried chicken dinner as well as the black poultry processors and cooks who prepared the meal (Williams-Forson 2006, 72).

Vigorous propaganda collaboration between the military and the poultry industry attempted to inform civilians about the value of keeping chickens. "Keep Poultry in Your Backyard/ In Time of Peace a Profitable Recreation/ In Time of War a Patriotic Duty" reads a poster issued by the United States Department of Agriculture; "Two hens in the backyard for each person in the house will keep a family in fresh eggs."[11] The Department of Agriculture issued new guidelines for supported poultry prices, and the *National Poultry Digest* called for "A Million Eggs for a Million Soldiers," observing that "this use by the Army of a million or more eggs a day means an increased demand for eggs all over the country. Food and eggs become one of the most important features for building up our defense ration, and at the same time our poultry and farm industry."[12] The effect on the poultry industry was dramatic: broiler production increased almost threefold; the vertical integration of the industry combined feed processing, hatcheries, grow-out facilities (both contract-based), and poultry processors into one company; and with the new economies of scale came new moves to rationalize and streamline the process of chicken raising. The American chicken became an assembly line product, made bigger, better, and faster—and with greater efficiency—for a public whose appetite was ever increasing, coaxed along by advertising and public relations (Striffler 2005, 46).

The Chicken-of-Tomorrow Contest

A public relations initiative between 1946 and 1961 known as the "Chicken-of-Tomorrow" contest reveals just how this new appetite was created and satisfied. This collaboration by American farmers, breeders, grocers, and agriculturalists was designed to reengineer the American chicken. Once again, film was central to this act of egg (and chicken) engineering. As film theorist Akira Lippit argues, film acts as a potent multiplier of the energy inherent in life itself. Film usurps the energy of biology for social processes, as instanced by the film images of animals that proliferated in the late nineteenth and early twentieth centuries just as material animals were fading from view. From its very beginnings in Eadweard Muybridge's studies of *Animals in Motion* (1899), according to Lippit, film technology captured the vitality of moving creatures. This parasitic appropriation of animal energy enabled human beings to create and fuel "the new, industrial environment" (Lippit 2002, 124).

The filmic practices of reshaping space and recalibrating time that had been central to the tissue culture laboratory also proved central to agricultural efforts to create an ideal chicken. In the late 1930s and early 1940s, agriculture began to focus on creating "meat birds": chickens raised specifically for slaughter and sale to the national market. Before that era, the main agricultural focus had been on egg production. Laying hens were kept specifically for that purpose, and the spent hens (those who no longer laid eggs) and roosters provided the family chicken dinner (Sawyer 1971, 26, 36, 112). That began to change in 1946, when Howard C. Pierce, specialist in poultry marketing for the A&P Food Stores, persuaded the American poultry industry to join him in launching a contest for small farmers and large commercial breeders to create what they called the Chicken-of-Tomorrow.

A bold new public relations venture, the contest ran nationwide in three-year cycles from 1946 through 1961. Regional contests in the first and second years culminated in the big national contest every third year. In addition to sponsoring the competition, the A&P Food Stores also contributed a cash prize of five thousand dollars to be awarded to the farmer or poultry breeder who produced the winning chicken.[13] Even the runners-up benefited, for they were permitted to advertise their birds as "Chicken-of-Tomorrow" contest finalists, a designation with welcome implications for their profit margins. For the first contest, in Georgetown, Delaware, in 1946, breeders shipped their entries in the form of fertile eggs, to be hatched, grown out under controlled conditions, and subjected to state and national rounds of expert judging (Sawyer 1971).[14] This was a fraught process, as the *Saturday Evening Post* described it (in a rather gendered formulation): "Each entry will be represented by 400 birds, which will give the judges the monumental chore of safeguarding and checking as many as 20,000 temperamental, feathered waifs."[15]

Three ideas formed the foundation of the Chicken-of-Tomorrow contest: the belief that the ideal chicken was the meatiest bird producers could raise, for the lowest feed cost; the hope that science and industry could collaborate to produce this ideal chicken; and the faith that the powers of the media could be harnessed to support this project. Belief and hope were challenged seriously during the contest's run, but faith was not misplaced, for with media help the Chicken-of-Tomorrow competition reshaped not only the chickens we eat and the poultry industry that sells them to us but arguably even the physique of the American population (both women and men). Innovations in chicken farming, production, and marketing—in particular a shift to large-scale growing and value-added products (or from whole-bird sales to fast food servings of Chicken McNuggets)—modeled the high-volume industrial livestock production that has given the United States inexpensive and uniform meat, accelerated the demise of the local farm, and caused—many argue—a raging obesity epidemic.[16]

An intricate collaboration lay behind what may seem merely a biological initiative to standardize American chicken breeding. Elaborate planning occurred at both local and national levels, so that (for example) the 1951 Chicken-of-Tomorrow Contest national planning committee included representatives from the International Baby Chick Association, the Poultry & Egg National Board, the Poultry Branch of the USDA, the poultry magazines and newspapers, the American Farm Bureau, the American Feed Manufacturers Association, and the Animal Husbandry department of the USDA along with one extension specialist in poultry husbandry and one professor of poultry science (from Ohio State University). State committees for the same year included "people representing various segments of the industry, such as a broiler grower, a breeder of meat-type poultry, a hatchery operator, a poultry processor, a wholesale or retail distributor, and [significantly] an editor or radio publicity agent."[17]

In retrospect, it was no doubt naïve to think that among this diverse group of stakeholders there could ever be agreement over the qualities of an ideal meat-type chicken. However, at the outset, representatives of the USDA approached academic poultry scientists asking for their assistance. Those poultry science professors were a tough lot to shepherd toward industry goals. Although the academic discipline of poultry science has had a warm relationship with poultry farming over the years, with many of its graduates working as extension agents, advisors to poultry companies, or in-house scientists, in its early days when it was busy establishing itself, the field emphasized the "science" part of its title over what preceded it: the production of poultry. Although industry representatives claimed the ideal chicken as both production goal and marketing strategy, poultry scientists were far more cautious. So when Melvin Buster, chief of Market Standards and Facilities at the USDA, approached Dr. I. M. Lerner of the poultry department at UC Berkeley asking for help, Lerner responded immediately with serious objections to the project, "Since I have no photographs, or even a mental picture for that matter, of an ideal bird, I find myself in a position of being able to criticize but not contribute anything particularly constructive" (Keilholz 1946).[18] Professor Lerner went on to tactfully distinguish between questions amenable to scientific answers and those falling more in the realm of economics or social psychology:

I have done some work in the field of meat production, but somehow or other the development of an "ideal" meat type has not fallen within the purlieu of this work. The great difficulty in my mind is the fact that neither the heritability nor the criteria of what the ideal attributes are have been worked out. . . . That is why I find it difficult to advance any more concrete suggestions. . . . Only when more precise information on (1) the contribution of each factor to the gross value of the bird and (2) the heritability of each factor which contributes significantly to this gross value, is available can anything more than a purely subjective score card be worked out. A combination of many subjective score cards will not in my opinion result in an objective one.[19]

Butler's response to Lerner made the USDA's agenda clear, advising that to the committee it seemed clear that "the Cornish x New Hampshire F_2 cross produces the best carcass considering rate of growth, percentage of meat to bone, and amount of breast meat which appears to be an important consumer preference factor." He closed with the hope that "a few state college poultry departments will . . . furnish the committee (1) photographs, front and side views, of dressed carcasses of their selection of individuals of both sexes and of different ages . . . and (2) two or three wax or plaster cast models of the best dressed carcasses."[20] Notwithstanding Butler's request, however, Lerner and his fellow poultry professors refused to help the contest by itemizing favorable traits, writing the scorecard, or sending the contest committee photographs or plaster casts of so-called ideal chickens.

Where collaboration was not to be found, its counterfeit could be manufactured. It seems likely that Carl Byoir and Associates, the New York public relations counsel for A&P Foods, simply selected phrases from correspondence and used them out of context in their press releases. Thus, an article offering a "Preview of Tomorrow's Chicken" seemed to endorse the project when it quoted "I. Michael Lerner, of the University of California College of Agriculture at Berkeley. . . . [who] sees tomorrow's chicken as the answer to these questions: What characters are economically significant, how may these characters be measured, to what extent are they heritable, how much does each character contribute to the market value of the bird?" (Keilholz 1946).[21] In the end, the hoped-for collaboration between poultry science and the food and poultry corporations dwindled to a pro forma reporting relationship. The poultry science professors were merely "kept informed" of the committee's progress in generating the description of the ideal chicken.[22] It is scarcely surprising that both the 1948 and 1951 contests were won by a Cornish Rock cross-hybrid, just the type of bird that Buster thought might be ideal.[23]

From the contest's beginnings as a public relations scheme, media was integral to its various initiatives. Although the ground rules for contest judging clearly mixed breeding standards with marketing priorities in a rather unscientific fashion, newspaper and magazine stories frequently presented them without comment or critique. Thus the same article in the *Saturday Evening Post* that had gushed over the monumental chore of safeguarding "as many as 20,000 temperamental, feathered waifs" reported that "specifications called for a bird that puts on weight rapidly and thus at less cost" (Sawyer 1971, 116). It went on to cite Dewey Termohlen of the USDA: "I believe that within five years as a result of the Chicken-of-Tomorrow program, we will see a complete revolution in the production and marketing of poultry meat. It is the first real demonstration of *production aimed at marketing*" (Sawyer 1971, 117; emphasis added).

Indeed, as the contest continued, the ideal chicken's qualities were not so much scientifically determined as they were commercially generated, keyed to two industry goals: speedy production and market appeal. Chickens were now being bred to meet production demand for a well-fleshed, well-proportioned, but above all quickly growing bird.[24] Chickens were grown increasingly not in backyards, where food of all sorts was plentiful, but in poultry houses, where food had to be supplied. Therefore, the shorter a chicken lived and had to be fed, the better. Not merely the amount of meat but its type and color as well were assessed by both judges and the consuming public, although there is little explicit verbal acknowledgment of this more culturally laden set of criteria. Instead, the color preference went unstated, and images did the work. For example, a semiotically overdetermined publicity photograph for the Chicken-of-Tomorrow contest shows a young white woman gazing at scales that hold what seems to be the meat from two chicken carcasses. The left-hand scale rides high, containing a small pile of predominately dark meat, while the right-hand scale dips under a larger pile of what seems to be only white meat. The caption reads, "HOUSEWIFE SEES DIFFERENCE in average market-type chicken and meat-type bird from contest. Both birds were same age. Average bird at left had less meat and more bone than Chicken-of-Tomorrow at right."[25]

The production of standards for gender and race are in evidence right alongside more straightforward manipulation of biological life in the archives of the National Agricultural Library that document the history of U.S. poultry breeding. In agricultural fairs repurposed as Chicken-of-Tomorrow fairs, poultry breeding contests were accompanied by their eugenic cultural cousins, beauty contests. To be sure, women were eligible to compete as poultry breeders. From the beginning they appeared on the list of winners cloaked in their husbands' names or marching boldly under their own: Mrs. Sam Smith of Eureka, Illinois, with her New Hampshires, and Mrs. Elsie Jacobs with her White Plymouth Rocks.[26] However, women were far more evident among the winning beauties than the winning breeders, for reasons that become apparent when we consider two of the very few pictures of women to be found among the many

FIG. 4. Prize hens and beauty queens.

surviving photographs of the (predominantly male) contest winners. The first photograph, picturing a young girl in a dark jumper receiving a prize from a dark-suited man, is captioned "Mary Bernard of Natchitoches Parish 4-H Club member whose pen of 6 White Rock broilers was named Reserve Champion of the 11th Annual Chicken-of-Tommorrow [sic] . . . with C. L. 'Spec' Varnado of Natchitoches who paid a premium of $150.00 for the birds for Ralston Purina Co., Jackson, Mississippi."[27] Behind the winning girl, past the large placard of a white chicken, we glimpse a ribboned stage on which are ranked two rows of beauty contestants. On its reverse side, the photograph is marked, in pencil: "DO NOT USE." The second photo has the same setting, but now we see a brown-haired woman in a white ball gown, with a sash that seems to read "Vermillion." She has also accepted a trophy from a man in a gray suit. On the back, in pencil, the photograph is labeled, "La. Poultry Queen."[28] Clearly greater publicity value lay in the search for a "Chicken-of-Tomorrow" queen than in a winning woman chicken breeder.

Both the beauty queen and the prizewinning chicken must have a pleasing physical appearance, as one box in the archives inadvertently reminds us. We find there the judging specifications for the Miss Delmarva Contest, part of the Delmarva Chicken-of-Tomorrow Festival of June of 1948: "The Queen will be selected on the basis of Beauty, Posture, Poise and Personality. . . . [she] will receive a total of $1000 in valuable awards."[29] Recall that the breeder whose bird won "Chicken-of-Tomorrow" was promised five times that amount. In the same box we find a set of red plastic calipers, labeled "Poultry-Breast Meter for Measuring Breast Conformation of Meat-Type Poultry." The juxtaposition of these archived objects is eloquent, if unintentional.[30]

I have suggested that the poultry industry sustained this resemblance between chickens and women because it produced good copy. A March 1951 article by Tom Wilkinson in the *Arkansas Agriculturalist* suggests another reason: the increasing cultural cachet of Hollywood. Wilkinson proclaims, "The day of the slick-hipped chick is over. Now it's the era of the pleasantly plump, balloon-breasted model built not on the lines suggested by Paris and New York fashion designers, but by the leaders of the Chicken-of-Tomorrow Program." The article goes on to imply that the magic of Hollywood may even be behind the superior physique and performance of the new hybrid chicken. "Yes, a remarkable change of appearance in chickens has occurred in the past five years. Hollywood script writers might describe it as 'wondrous' and 'supercolossal' because it has transpired in less than six years, half the possible normal life span of a chicken." Wilkinson even suggests that American chickens have undergone a movie-star-like transformation, from "slender, slim-picking birds" to "today's meat-packed numbers" (Wilkinson 1951).

In its recourse to Hollywood movie language to explain a phenomenon in agricultural history—the speed at which, through hybridizing, chickens have been transformed into much larger birds—this article also shows how central film has become to the modern reshaping of life (Landecker 2007). We have seen the essential role of film in the biological study of embryonic development, dating back to the collaborations in photomicrography of Waddington, Fell, and Canti, whose time-lapse records of developing chick embryos helped them acquire research funding and created a broad public interest in embryology. We have seen how film fueled the improvements in poultry breeding that were the goal of the Chicken-of-Tomorrow contests. And we can see how film played a crucial role in the development of mass market food products if we consider an event held in New York City on February 3, 1949.[31]

"Chicken Booster Day," a joint venture of the Poultry and Egg National Board, the Northeastern Poultry Producer's Council, the Chicken-of-Tomorrow Committee, Twentieth Century–Fox Film Corporation, and the Roxy Theater, featured a lavish dinner at the Hotel Statler followed by a private screening at the Roxy Theater of *Chicken Every Sunday*, starring Celeste Holm, Dan Daily, and the eleven-year-old starlet-in-training Natalie Wood.[32] During the after-dinner speeches at the Hotel Statler, event organizers also announced a "Chicken Every Sunday" contest. Entrants were asked to "write an essay, a limerick or a slogan of 100 words or less on 'Why I like chicken every Sunday,'" which they would send Fred Smith, eastern manager of the Poultry and Egg National Board, including "a real chicken wishbone" with each entry. "Three regional areas, East, Central, and West, will each wind up with one winner, who will be taken to Hollywood, expenses paid, as a three-day guest of 20th Century and Dan Daily and Celeste Holm, stars of the *Chicken* film. From the three winners one national winner will be chosen at a chicken dinner celebration and he will be given his wish—unless it exceeds $1,000 in cost, in which case he is handed over the $1,000 in cash."[33] Even the chickens served at the Chicken Booster Day dinner were media-and-industry engineered starlets, according to the *Herald Tribune*:

> BARNYARD QUEENS—The banquet birds were from one of the prize-winning flocks participating in the Great Atlantic and Pacific's Chicken-of-Tomorrow Contest . . . [which] has as its goal a broiler weighing in at least a full pound more than the average broiler of today, and coming to this hefty size on no more food and in far less time. The broilers served at the party were from the prize-winning flock of Widehamps, a pure strain of New Hampshire developed by J. E. Weidlich, of Roanoke, VA. *These sweater girls of the barnyard were ten weeks of age and weighed more than three pounds apiece before dressing.*[34]

This link between film and industry through the *Chicken Every Sunday* screening, like the writing contest announced at the Chicken Booster Day dinner and the entire system of Chicken-of-Tomorrow contests, furthered the rise of the new field of industrial agriculture. It also exemplified the new marketing strategy that would within a decade change the grocery business in the United States as regional consolidation led to national dominance in the grocery industry and then an antitrust suit against the A&P Food Company. Savvy in their use of print and film media, the A&P Food Company used public relations projects to persuade poultry producers to adopt uniform poultry breeding as part of a far-reaching industrial food production and distribution system.

Unlike the excitement produced by the embryological films of the 1930s or the futurological sound of a "Chicken-of-Tomorrow Contest," however, this particular push to redirect the biological future of poultry was sold to the public by way of the past. Egg, meet advertising: another flexible medium, able to incorporate contradictory desires and impulses. The nostalgic chicken dinner eaten in *Chicken Every Sunday* could not have been further from the poultry products resulting from the innovations in hybrid poultry breeding catalyzed by the Chicken-of-Tomorrow Contest. The biological makeup of the chicken population of the United States changed, thanks in part to the Chicken-of-Tomorrow contests and their spin-off, the Chicken Booster Day, and so did "the genetics of the broiler industry," according to one poultry historian. "In a six-year span the entire emphasis on broiler breeding was changed and intensified. In addition, research stepped up in nutrition and feeds, management, medication, and other fields. The technical impact was both immeasurable and immense" (Sawyer 1971, 123). These public relations events propelled American chicken breeders into reengineering their product, thus manipulating their breeding stock to produce chickens with the specific physiology and accelerated growth rate that would meet the demand of poultry producers.

The final step in this transformation of American chicken was to change its image in the eye of the consumer from a special once-a-week poultry meal into an everyday meat. Moving away from the butcher shop offering whole hens, processors now produced and marketed chicken products in a range of forms and sizes: from chicken parts sold in uniform packages of thighs, legs, or breasts to specially formed chicken patties, and finally chicken hot dogs, chicken strips, and chicken fingers (Horowitz 2006). As geneticist Guy Barbato points out, "The domestic and international poultry industries have gone through many changes over the last fifty [years]. The industry has matured from a 'backyard flock' mentality, supplying Sunday chicken dinner, to sophisticated mass production of poultry meat and eggs" (Barbato 1991, 444).

The chicken of today is a product of cooperation and collaboration between biological and embryological researchers, the universities, the Cooperative Extension Services and 4-H Clubs, the emerging fields of advertising, public relations, and marketing and the new vertically integrated grocery and poultry industries. Most of all, however, the chicken of today is the result of a provocative new biology dedicated to intervention in individuals and populations in space and time. The new biological interest in engineering life links those chick embryo experiments in the 1930s to the new strategy of remaking the poultry population embraced by mid- to late-twentieth century American chicken breeding. And—a fact worthy of metaphysical speculations—it links the chicken of today to the research on human embryos debated by the British Parliament in 1989, when those British Egg Board advertisements were urging people "go to work on an egg."

Culture

"Culture in all its early uses was a noun of process: the tending of something, basically crops or animals."

–Raymond Williams, *Keywords*

From the large-format photographs of colorful chickens decorating the restaurant on Tenth Avenue where my family celebrated Father's Day to the smashing
close-ups of exotic breeds in a calendar I leafed through at the Barnes and
Noble on Sixteenth Street, New York City seemed full of chicken culture in
2007.[1] What do I mean by that term? Until very recently I would have meant
something agricultural. As Raymond Williams has observed, "culture in all its
early uses was a noun of process: the tending of something, basically crops or
animals" (Williams 1983, 87). Yet, whereas chicken raising is becoming a greater
presence in the five boroughs every day, chicken culture stretches beyond its
original meaning—the cultivation or tending of chickens—to refer to an
increasingly widespread current phenomenon: the production and exhibition
of artistic works that incorporate chickens to generate a wide range of aesthetic,
social, and scientific meanings.[2]

The notion of chicken culture draws together two seemingly disparate
worlds: the metropolitan art market, where values are established by critics
and galleries, and the agricultural market, where values are set by the craft and
science of animal breeding. Common to both worlds is one specific kind of
visual image: the chicken portrait. This type of image—initially an engraving or
painting and now frequently a photograph—was originally produced to help
the practical poultry manager identify specific chicken breeds and to provide
the poultry fancier with a "standard of perfection" to which he might aspire as
he bred exhibition birds. From its heyday in Victorian England and America to
its reinvention in the works of contemporary photographers, sculptors, and
conceptual artists, the chicken portrait has served as a powerful instrument
for producing both chickens and the human beings who tend or attend to
them.

Poultry Shows and Hen Fever

Chicken images first proliferated noticeably in the United States in the mid-nineteenth century. A craze for poultry breeding had been fueled by a new social phenomenon, the competitive poultry exhibition, where poultry breeders could win increasingly lavish purses for their prize birds. The first major poultry show in the United States was held at the Public Garden, Boston, November 15–16, 1849. Beginning in 1890, when the Madison Square Garden poultry show became an annual event, poultry shows were permanent features at state agricultural fairs throughout the United States, and poultry exhibition buildings rose up as part of the grand exhibitions of the era: the Chicago World's Fair in 1893, the Pan-American Exposition at Buffalo in 1902, the 1904 St. Louis Exposition, the 1907 Jamestown Exposition, and the 1915 Panama-Pacific Exposition in San Francisco (Robinson 1921b, 7, 14).

As interest in exotic chickens increased, so too did another source of potential profit: the publication of illustrated poultry books to feed the appetite of poultry breeders and exhibitors. John H. Robinson's 1921 volume, *Standard Poultry for Exhibition*, is a classic example, with its extensive and exhaustive table of contents covering judging fundamentals, the ethics of readying birds for exhibition, care of the exhibition stock, selection and fitting of birds for exhibition, care of birds in transit and at shows, and the method and philosophy of poultry judging.

Robinson's volume pictures the basic practices of chicken exhibiting: teaching the birds to pose, washing them, drying their feathers, and even bleaching or bending their feathers to the proper angle. *Standard Poultry for Exhibition* is notable for the marvelous compendium of photographic portraits it includes, both individual and group. We see standard and inferior chickens organized by breed and sex, and we see serious dark-suited poultry exhibitors, poultry show organizers, and poultry judges, nearly all uniformly white and male.

Dedicated to illustrating the proper and improper physical type of each major chicken breed, the book develops in a reader the habit of making morphological distinctions, or approaching the human portraits with a eugenic gaze. Thus, looking at the portraits of "Prominent Black Minorca Breeders and Exhibitors" and "Prominent Breeders and Exhibitors of Asiatics," one is tempted to begin judging the genetic information conveyed by the features of the men pictured—to note the pince-nez, the glasses, the prominent ears, the drooping moustaches and beards, the receding hairlines, and the firm or receding jaws among those orderly images of white men. When we have examined the portraits of prizewinning and faulty model heads of single-comb males and single-comb females, all pictured together in three rows of four portraits on facing pages, it seems inevitable that we will make the same close examination of the

FIG. 5. Prominent breeders and exhibitors of Asiatics.

prominent women exhibitors, pictured in four rows of four, in the final chapter, "Logical Developments in Judging Practice," a mere four pages from the end of the book.

This craze for poultry exhibiting and the resulting appearance of the new genre of poultry books appears with acerbic commentary in a classic mid-nineteenth-century satire, *A Short History of the Hen Fever* (1855). As George Burnham recounts, "hen fever" began when he received an order of the new

FIG. 6. Model heads of single comb males.

"'*Cochin*-China' fowl, then creating considerable stir among fanciers in Great Britain," which came "direct from Queen Victoria's samples" (16). The craze for poultry exhibits had begun, first with the show in the Boston Public Garden and next in New York City and points beyond. That initial Boston poultry show awakened an economic and emotional enthusiasm for poultry exhibiting that took the whole country by storm: "The exhibition lasted three days. Unheard-of prices were asked, and readily paid, for all sorts of fowls. . . . As high as thirteen dollars was paid by one man (who soon afterwards became an inmate of a lunatic asylum) for a single pair of domestic fowls" (24). For the first time, town dwellers were the ones swelling the booths: the exhibitors were "city men with whom farming was more or less a hobby" to which they gravitated more for the pleasure in competition than for the profit. This new metropolitan zest for chicken breeding and exhibiting launched a wildly inflated market for exotic birds. Drawing on the term for inflationary spirals in economics, Burnham calls this "Bubble Number One." "Never in the history of modern 'bubbles,' probably, did any mania exceed in ridiculousness or ludicrousness, or in the number of its victims surpass this inexplicable humbug, the 'hen fever'" (21). An excited press further fueled the upward spiral of prices, writing of "'extraordinary pul-

FIG. 7. Model heads of single comb females.

lets,' 'enormous eggs,' . . . 'astounding prices' obtained for individual specimens of rare poultry; and all sorts of people, of every trade and profession and calling in life . . . on the *qui vive*" (21).

Burnham describes a similar upsurge of activity in the publishing world in *A Short History of the Hen Fever*, as poultry book writers respond to the new opportunity for windfall profits. Speed is of the essence in producing books, just as it is in producing hens, and Burnham knows this. When he learns that another poultry breeder has a book on chickens about to appear, he hires a stable of workers and sets them to the task of rush producing his own just-in-time publication:

> I engaged three engravers, who worked day and night up on the drawings and transfers of the fowls for my illustrations; the paper was wet down on Monday and Tuesday; I read the final revised proof of my work on Wednesday night; the book went to press on Thursday; the binders were ready for it as it came up, the covers were put on on Friday morning, and I sent to the New York house (who had bespoken them), by Harnden's express, on Friday evening, three thousand five hundred copies of the

FIG. 8. Prominent women exhibitors.

"New England Poultry-Breeder," *illustrated with twenty-five correct engravings of my choice*, magnificent, superb, unapproachable, pure-bred fowls. This book had an extraordinary sale,—far beyond my own calculations, certainly.   (30–31)

Fueled by these how-to guides, the sales of poultry increased, the claims of creating new breeds became more and more outrageous, the prices paid for such "purebred" creations less tethered to any realistic value, and the whole enterprise took on the character of a circus, Burnham tells us. Even famed impresario P. T. Barnum caught the fever, becoming a "Hen-man" and taking on the presidency of the National Poultry Society (194). Eventually the "leading dealers in fancy poultry" came to include "clergymen, doctors, and other 'liberally-educated' gentlemen" (155). These were drawn into the boom by the opportunity to speculate in a rising market, Burnham explains: "The fever for the 'fancy' stock broke out at a time when money was plenty, and when there was no other speculation rife in which every one, almost, could easily participate. The prices for fowls increased with astonishing rapidity. The whole community rushed into the breeding of poultry, without the slightest consideration, and the mania was by no means confined to any particular class of individuals . . . our first men, at home and abroad, were soon deeply and riotously engaged in the subject of henology" (305).

As with other such economic bubbles, the hen fever bubble eventually burst, leaving even wealthy investors bankrupt as potential buyers moved on to Merino mania.[3] Yet the economic cost of the craze for exhibiting exotic chicken breeds was unmatched in its day. "I have asserted," Burnham tells his readers, "that, in all probability, in *no* bubble, short of the famous 'South Sea Expedition,' has there ever been so great an amount of money squandered, from first to last, as in the chicken-trade" (294). Yet the money may not have been squandered, exactly; Burnham's satire also predicts a new economic phenomenon: poultry breeders' business profits would soon be determined by their success at poultry exhibitions:

> [The] money value of the birds is determined by their desirability to people who are able to indulge their desire to have the best of everything who take an interest in fine fowls. This class of people may be divided as to their interest in poultry into those who will pay liberally for fine specimens, especially those with a prize record, simply as ornaments for their ground or to satisfy their ambition to have the best; and those who consider either wholly or in part the breeding value of the birds. These two groups bid against each other and against the professional breeder who can often afford to pay a very high price for a bird either to use in his own yards or to keep it out of a competitor's hands.
>
> (Robinson 1921b, 3)

The hen fever that began in the United States with Burnham's imported Cochin-China hens from Queen Victoria's best stock also flourished in the United Kingdom in that era. As in the United States, where Robinson's *Poultry for Exhibition* provided information to gentlemen and some ladies eager to participate in poultry shows, in Great Britain the new genre of advice books for the poultry breeder was also gaining more space and even defining a new segment of the publishers' lists. Most books in this genre offered finely drawn colored illustrations as an index to the ideal breed standards, offering as a side benefit to readers a range of rich visual and aesthetic pleasures by documenting ever more beautifully feathered novel breeds (Sayer 2007, 593).

Among the poultry breeder's handbooks that appeared during this era of hen fever on both sides of the Atlantic, one book stood out not only for its thirty chromolithograph color plates by the well-known engraver and animal portraitist Harrison Weir (1824–1906) but also for its dual-purpose perspective. As the author explained, the book aimed "to combine ornithological research with practical experience in . . . management" (Sayer 2007, 590). British ornithologist and naturalist William Bernard Tegetmeier, who appeared with Alfred Russel Wallace on the list of Extraordinary Members of the British Ornithologists' Union and would later become a trusted source of Charles Darwin, published *The Poultry Book* serially between 1866 and 1867, concluding with a new and illustrated edition in 1873 (590).[4] While aimed primarily at poultry breeders, *The Poultry Book* and volumes like it encouraged communication between the realms of literature, art, and science that paradoxically gave rise to two seemingly contradictory impulses: "to the protection of birds in the wild . . . and to their increasingly scientific management in agriculture" (589).

Weir's poultry portraits, like Robinson's later *Standard Poultry for Exhibition* and the all-important *American Standard of Perfection* (Drevenstedt 1898), participated in the convention of animal illustration famously established by Robert Bakewell.[5] Like his works, these volumes too can be said to embody the taxonomic gaze. They carefully categorize and portray the ideal type of each different breed, following conventions not of fine art but of agricultural shows, even enhancing the animals' most highly valued features to aid the farmer-breeder in identifying those qualities for which they should aim (Sayer 2007).

"On the Principal Modern Breeds of the Domestic Fowl," an essay Tegetmeier published in *Ibis* in 1890, exemplifies the focus on the poultry breeder and exhibitor that he shared with Weir, and it anticipates the interest in avian (and egg) engineering that would come to dominate poultry breeding in the next century. A bird's value can be increased, the article stresses, by incorporating some desired traits, yet that act is risky, since by such modification the breeder could also cause other highly valued traits to disappear (Tegetmeier 1890). For example, he writes, the Black-breasted Red Game cockerel and pullet have shown a dramatic change "since the establishment of competitive poultry shows, [as] the breeders have aimed at increasing the length of the neck and limbs" (305). While observing that the object of selective breeding was "to produce breeds that should excel others in the conventional fancy points, so as to be able to win prizes at competitive shows," Tegetmeier explains that "exhibition specimens having been bred so exclusively for show purposes, no attention having been paid to breeding for egg-production, . . . they have lost this characteristic and are now very indifferent layers" (310). Here Tegetmeier shows an awareness of the unanticipated negative effects of breed modification that later poultry breeders would lack, as the weight-bearing deficiencies of the Cobb 700s so dramatically demonstrate.

While Tegetmeier's entire survey of domestic poultry elegantly combines beautiful engravings of a range of distinct breeds from the Black-breasted Red Game Cockerel and the Silver-Spangled Hamburg Cockerel to the outrageously coiffed Spangled Polish Hen and the elegantly rounded Wyandotte Hen, it is most notable for the history it tells of the rise of chicken exhibiting in the United Kingdom. "The introduction of the Cochin . . . gave the first impulse to the exhibition of fancy poultry and to the manufacture of new breeds by cross-ing different varieties," Tegetmeier tells us, and he notes that most recently the essentially "plastic character" of the species *gallus* has led poultry fanciers to attempt to produce Bantam varieties of the most familiar breeds. "This has been most successfully accompanied with the Game, the Malay [and] the Cochin" (Tegetmeier 1890, 326, 327). We have here another example of vast dif-ference in scales on which avian life has been engineered: two decades before biologists and embryologists began to use cinematography and tissue culture to exploit the plasticity of the avian egg and embryo, and a century before the poultry industry claimed to have created the "World's Most Efficient Chicken," the Cobb 700, chicken fanciers and agriculturalists were drawing on the same valued traits working at the scale of the breeder's own flock.

As one of the dominant authorities in British poultry management and breeding, Tegetmeier "participated in the wider cultural shift that captured the wild for tamed nature as understood through human culture: preservation, recreation, study, and exploitation" (Sayer 2007, 601). His work was a major influence on Charles Darwin's 1868 study, *The Variation of Animals and Plants Under Domestication*. In a footnote to the chapter on "Fowls," Darwin explains, "I have drawn up this brief synopsis from various sources, but chiefly from information given me by Mr. Tegetmeier, [who] has kindly looked through this chapter . . . and has likewise assisted me in every possible way in obtaining for me information and specimens" (Darwin 1858, 232). Tegetmeier and Darwin brought together a combination of naturalist observation and breed specifica-tions in the service of a closely reasoned argument in support of the theory of "unconscious selection": the idea, introduced earlier in *Origin*, that poultry breeders may unintentionally modify their breeds simply by preserving the best birds and destroying the lesser ones. "It is not pretended that any one in ancient times intended to form a new breed, or to modify an old breed accord-ing to some ideal standard of excellence. He who cared for poultry would merely wish to obtain, and afterwards to rear, the best birds which he could; but this occasional preservation of the best birds would in the course of time modify the breed, as surely, though by no means as rapidly, as does methodical selection at the present day" (Darwin 1868, 240). "Fowls" locates the ancient roots of deliberate breed improvement in the simple practice of taking care of valued chickens, and it identifies the "parent-source" of all breeds of chicken: *Gallus bankiva*, the wild Indian chicken (Darwin 1868, 242).[6] The practice of breeding fancy poultry for competitive poultry exhibitions, such as Darwin's naturalist project, exemplifies the Victorian appropriation of the natural world for the good of the empire (Sayer 2007, 601).

The Sweet Chick

We have seen how the transition from chicken culture as an agricultural phenomenon to a chicken culture also incorporating aesthetic, social, and economic values began with the rise of poultry exhibiting and the allied market for poultry books. But how has this new phenomenon had an impact on metropolitan culture more generally? American humorist S. J. Perelman's short story "The Sweet Chick Gone," published in his volume of satiric tales *Chicken Inspector No. 23* (1952), marks and mocks the craze for fancy poultry. Like Burnham's *A Short History of the Hen Fever* a century before, Perelman's tale dismisses the fad as one more "daffy idea," like the notion there is "big money in chinchillas." Yet the theme of agricultural obsession does not keep this chicken story from taking a characteristically modern narrative twist. Rather than being told by a noted chicken breeder as in Burnham's *Hen Fever*, Perelman's story is recounted by the "Sweet Chick" herself: a White Leghorn who has been purchased from a hardware store in Perkasie, Pennsylvania, by the suburban householder (and the author's comic alter-ego) S. G. Prebleman. Far from a cherished exhibition chicken, this Leghorn begins her story lodged in a "sleazy coop" made by her owner who knows "nothing whatever about rearing domestic fowls." Rather than the "traditional diet of mash and cracked corn," she is fed "potato chips, olives and soggy cashews, stale hors d'oeuvres, soufflés that had failed to rise, and similar leftovers" (Perelman 1952, 132). Still, outliving her eleven comrades, who "croaked almost at once," the sweet chick goes on to attain "a special niche in Prebleman's affections. Whenever he trundled a barrowfull of empty bottles to the dump (a daily ritual, it seemed) he would stop by to gloat over [her] progress" (132). Finally, one spring day, she learns what that affection means:

> "Hi ya, gorgeous," he cooed, smacking his lips. "Mm-hmm, I can just see you in a platter of gravy next week, with those juicy dumplings swimming around you. Hot ziggety!"
>
> Horrified at the doom in store, snatching at the only means I had to avert it, I . . . recovered my faculty of speech. "You-you vampire!" I spat at him. "So this is why you coddled me—fattened me—"
>
> "What did you say?" he quavered, turning deathly pale.

"Not half of what I'm going to!" I said distractedly. "Wait till I spill my story to *Confidential* and the tabloids. . . ."

He stuffed his fingers into his ears and backed away. "Help! Help!" he screamed. "It spoke—I heard it! A talking chicken!"   (132)

Prebleman quickly recovers from the surprise at finding himself the owner of "a glib chick" when he realizes he could reap huge economic benefits from his possession. They should collaborate on the story of her life, he urges her: "Three days after this book of ours appears, it'll lead the best-seller list. You'll have guest shots on Paar, Sullivan, Como—you pick 'em. The magazine rights alone'll net us three hundred G's. Add to that the movie sale, the dramatization, and, of course, the diary you keep meanwhile, which George Abbot turns into a musical so the whole shooting match starts all over again. No more white-washed roosts for you, girlie. You'll be sleeping between percale sheets!" (132)

Bewitched by his assurances, the sweet chick dreams of "sporting diamonds the size of hen's eggs" (133). At first, those dreams really do seem to come true. The publication of their coauthored chicken autobiography makes "literary history. Overnight, her memoir *Vokel Yolkel* became a sensation, a veritable prairie fire." The book goes through five editions in two weeks, she undertakes a fifty-city speaking tour, and, most notably, "David Susskind announced a four-way telecast between the two of him, myself, and Julian Huxley on the topic 'Which Came First?'" (134). Perelman's imaginative tour-de-force brings naturalist film and embryological cinema photomicrography together in a surreal television event: Julian Huxley, zoologist, embryologist, architect of the modern evolutionary synthesis, and producer of the Oscar-winning film *The Private Life of the Gannets* (1934), sparring with David Susskind, talent agent, producer, and famous talk show host, to solve the ancient riddle: Which came first, the chicken or the egg? (see Squier 1994, 140–143; and Mitman 1999, 78–81). Fame notwithstanding, the sweet chick is in trouble. Prebleman, her benefactor, has become her jailer, tucking her away "in a shabby theatrical hotel on New York's West Side . . . while he lolled in luxury at the Waldorf" (Perelman 1952, 134). She is finally released from her captivity by a talent agent named Wiseman, who promises that he can break her contract by challenging Prebleman with blackmail "for transporting a chicken across a state line." Wiseman spirits her away in a pillowcase to "to a prepared hideout in Fordham Road, a kosher butcher shop where [she] remained incognito in a flock of other Leghorns until litigation subsided" (ibid). When the lawsuit is resolved by a cash settlement from Prebleman to the lovely Leghorn, he is left a broken man, searching through "various obscure Pennsylvania hatcheries, pathetically mumbling to the crates of chicks as they emerged and hoping to duplicate his coup" (ibid).

S. J. Perelman's satire ends with the sweet chick ensconced in a suite at the Sherry-Netherland hotel, where she perches on the vanity and gazes into the mirror:

> While not given to excessive conceit—we White Leghorns, if less flamboyant than your Orpingtons and Wyandottes, take pride in our classic tailored lines—I was pleased by the reflection the mirror gave back to me. My opulent comb and wattles, the snowy plumage on my bosom, and the aristocratic elegance of each yellow leg adorned by its manicured claws all bespoke generations of breeding. In this great metropolis . . . was there, I asked myself, one pullet half so celebrated and successful, a single fowl who could contest my supremacy? . . . I had reached a pinnacle undreamed of in poultrydom, but at what fearful cost.   (129)

Although she is "the luckiest and loneliest creature on earth," she still misses Prebleman, "the most lovable galoot I had ever known": "I had everything—my own television program, a syndicated newspaper column, reserved seats at every preem, jewels, cars, furs—but it was so much tinsel. I would gladly yield it all to nuzzle the man's dressing gown again" (135).

While recalling the poultry craze of the previous century, Perelman's short story also claims a contemporary target: mid-century nature films whose tender narratives are far from the clinical objectivity of Canti and Waddington's tissue culture films or Huxley's gannet bird-blind observations: as Perelman puts it, "*Another Plum from the Anthropmorphic Pie Containing Born Free, Ring of Bright Water, etc.*" (135).[7] Half a century later, the short story also seems a nostalgic tribute to the arts and sciences of farming: to incubator-hatched chicks on sale as impulse items at hardware and feed stores; to the conviction that chicken raising could help the household economy; and, most of all, to the practices of poultry breeding and exhibiting, which rely on the fine points of breed identification that distinguish the blowsy, curvy Buff Orpingtons and Wyandottes from the trim White Leghorns.

Just as Perelman's "sweet chick" used an urban sensibility to rework the practices of rural poultry culture, a similar translation from breed styling to fashion and high culture appears in the work of contemporary photographer Jean Pagliuso. Although their milieu is high art rather than popular satire, the chicken portraits comprising "Poultry Suite" also explore issues of economic speculation, the nature of the market, and what might be called standards of perfection, for a breed or for a nation.

Poultry Suite

I first encountered Jean Pagliuso's work when an advertisement in the *New Yorker* drew me to the Marlborough Gallery in February 2006. There, in a suite of rooms upstairs from the exclusive restaurant Nobu, I found the collection of dramatic and visually stunning large-format, black-and-white photographs comprising a solo show, "Poultry Suite." I had come with a filmmaker friend who also keeps chickens. As we examined each photograph, we found ourselves alternately guessing the bird's breed ("This is a Spangled Homburg." "No, it's a Seabright!") and laughing at the resemblances we saw to famous movie stars. One very upright small white chicken pictured head-on with its legs tightly together reminded me of Charlie Chaplin, while the upswept profile of another, gazing off into the upper distance, evoked Greta Garbo. And then there were the feathers to marvel at, so close to couture dresses that the whole show reminded us of New York's Fashion Week.

Intrigued by the mixture of high art, fashion, and the characteristically rural practice of chicken exhibiting in "Poultry Suite," I arranged to meet Jean Pagliuso later that spring in her Chelsea studio, where the walls were lined with photographs of chickens. Pagliuso, who collaborated for a long time with film-maker Robert Altman, is known for her portraits of film stars, and the floor was strewn with celebrity photos.[8] In addition to Susan Sarandon, whose photo I glimpsed near my feet, she has photographed Robert Redford, Sam Shepard, Jack Nicholson, Meryl Streep, Keith Richards, and Sophia Loren. Pagliuso turned her attention away from celebrity and film culture in 1986, however, having recently retired from her thirty-year career as a commercial photographer,

FIG. 9. "Pagliuso White # 6 2005."

Used by permission of Jean Pagliuso.

FIG. 10. "Pagliuso Black # 15 2006."

Used by permission of Jean Pagliuso.

publishing in such magazines as *Vogue Italia, Harper's Bazaar, Newsweek,* and *Rolling Stone.* Visits to ancient sacred sites in Angkor Wat, Cambodia; India; Turkey; the California desert; and Peru resulted in the photographs comprising her first Marlborough Gallery show, "Fragile Remains."[9] The poultry photographs were her second venture into art photography.

Pagliuso describes "Poultry Suite" as homage to her Southern California childhood, when she assisted her father in the breeding and showing of Bantam Cochins—a miniature variety of the Cochin breed whose introduction into England "gave the first impulse to the exhibition of fancy poultry and to the manufacture of new breeds by crossing different varieties" (Tegetmeier 1890, 310). As the son of a Glendale ranch foreman turned citrus grower from Southern Italy, Pagliuso's father had chickens in his blood. He began raising them when he was seven years old, she told me: "Even in college, in the fraternity, he had chickens in the backyard, and he never stopped. So I was raised with him raising chickens. And then I started doing it too, and then we showed up in the Petaluma fair together, the state fair. And so it is a *huge* part of my life. My entire back yard was chickens. . . . And we lived . . . up in the foothills—but in the middle of town where you're not supposed to have that. And at four o'clock they were all up crowing."

A nationally known breeder and exhibitor, her father's success lay in his ability to wait generation after generation to breed birds with superior show

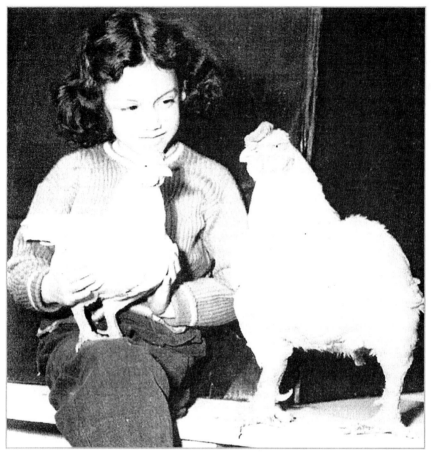

FIG. 11. Jean Pagliuso as chicken exhibitor.
Used by permission of Jean Pagliuso.

conformation: "We had only one kind—a Bantam Cochin—and we *only* had those, buff, black and white, and . . . my dad was grand champion over and over again, because genetically . . . he was willing to sit there for seven years, while . . . he would get the feathers on the feet from one chicken and the comb from another . . . and people would come from all over the United States and stay with us and take home eggs, or take home a chicken, so they could do the same thing." Petaluma, California, was a hub of chicken breeding in the first half of the twentieth century, and poultry shows were central to Pagliuso's girlhood: "My dad used to take me from the time I was four and I'd just have to be there all day long with him and sit with all those chickens." Yet in retrospect, she acknowledged, the activities of a chicken breeder-exhibitor were out of place in suburban Southern California of the 1940s:

My dad used to wash them in the kitchen sink, and then we had a hair dryer, big powerful hair dryer, and he used to bleach the white chickens with Clorox, just a little bit, I think everybody does. He'd just fluff them all up, and go and sleep there the whole time when he was showing, so that he could get up with the chickens in the morning and they were all ok. It was pretty bizarre, because that wasn't his life—he was a building contractor, and he was very involved with the UCLA sports program, helping them get football players and basketball players into UCLA, so he was very busy, but this was just a weekend hobby.

After her father's death, when Pagliuso returned to Southern California to dispose of his effects, she began thinking about those chicken exhibiting days. The childhood activity that once seemed strange and embarrassing—"I mean, it isn't *anybody* else's upbringing that I know of, and what a weird one, it's not even chickens in the chicken yard, it's just rows and rows of chickens"—suddenly seemed worth examination. "I got to thinking . . . I've got to photograph my dad's chickens . . . at least I'll photograph the remaining chickens, because I had never taken them seriously in any way, you know, and I thought well I'll just do one parting photograph."[10]

The impulse was in character: her mother had been a rose gardener, and Pagliuso had photographed "the last rose that she had blooming as she was dying." While as photographic subjects chickens may seem far from the elegance of roses, both projects drew on skills Pagliuso had used for years in fashion photography. "So I went online and I just kind of put in 'chicken rental' and up comes all of these familiar names from the fashion business. I used to use props all the time, [and] you can have anything . . . and they'll find it for you, and several of them did chickens. . . . they have a chicken wrangler that comes along, that breeds them, and she just kind of makes sure they're faced in the right direction, and when they run off the table she comes and gets them. It's like any other prop, live prop." Although her fashion career gave her the expertise needed to set up the first shoot, it may also have made it a disappointing experience. The chicken wrangler had underestimated the poultry knowledge of this Manhattan fashion photographer, and showed up with poor props: "This chicken was completely full of chicken shit, and she had the nerve to bring it in. And I said—these are not Bantams, Bantam Cochins. And she said yes they are! And I said, no, I know what they look like . . . I mean, where are the feathers on the feet, for instance? And she goes: I think they just had a hard ride in! And I said, Listen, I'm afraid you can't put one over on me, but I'm going to shoot anyway." Nevertheless, Pagliuso soon had a number of photographs that satisfied her, and she experimented with several kinds of paper and different methods for printing them. "I tried it on rice paper, and it just looked so much different, so much better . . . I tried regular silver gela-

FIG. 12. "Pagliuso Black # 5 2005."

Used by permission of Jean Pagliuso.

tin prints in large size, oversized, and it looked right to me. So then I was kind of off and running, 'cause once I saw the way they looked I just kept on renting chickens."

The exhibit that resulted balanced the aesthetic of fashion photography with the eye of a poultry breeder. Its vision unites the fashion photographer attentive to the beautiful move, the power of negative space, and the values of stateliness, graphic line, and elegance with the young girl taught to show chickens by her chicken-breeder father. "I think [chicken breeders] think nothing of just getting rid of the chickens that don't [show well], the real breeders are ruthless about not keeping about . . . [birds that won't show well]." The lessons her father taught her were crystallized in one memory: as a girl, she had "a chicken called Lulu who got to be mine, clearly only mine, because he wasn't ever going to the fair. He was not right—REJECT CHICKEN! NEXT!—and he had an amazing personality, he followed me everywhere, I'd go hiking and I'd turn around and there he'd be following me just like a dog. He was really cute. But one morning, day after Halloween . . . a coyote got them. I don't remember being terribly upset about it. I'm sure I must have been, but it didn't stick with me." Pagliuso's own preferences in "Poultry Suite" perhaps reflect that early experience. She told me that she prefers "the ordinary [photographs]," like the dirty little false Bantam Cochin, even though "the ones that people really like are the very stately graphic bird—people much more respond to . . . a kind of wayward elegance than they do . . . [to] the ordinary ones."

Markets, Breeds, and Types

The market famously drove the culture of exhibition poultry breeding, from its origins in mid-Victorian England and America to its decline in the late 1940s. A remarkably similar set of market-based decisions drives the creation of fine art photographs. Indeed, the decisions Pagliuso describes herself making in one of her photo shoots recall the self-assessment of S. J. Perelman's "sweet chick" as she perched on her vanity in the Sherry-Netherland: "we White Leghorns . . . take pride in our classic tailored lines. . . . My opulent comb and wattles, the snowy plumage on my bosom, and the aristocratic elegance of each yellow leg adorned by its manicured claws all bespoke generations of breeding." As Pagliuso explains: "There's no difference between the fashion model and a chicken as far as I'm concerned. You look for the beautiful move, and you look for the way it takes up the white space, and it's a sort of figure-and-ground thing, I used to do that with fashion models too. . . . not being a guy, and being into the girls, I used to look at different things when I shot, and it's kind of the same—feels exactly the same. And their feathers are the way they're dressed."

In its combined homage to chicken shows and fashion shows, "Poultry Suite" reveals some of the cultural forms linking prize-winning chickens and supermodels as well as the fancy poultry market and the fashion market. However, if the show delighted my friend and me with its blend of fowls, film, and fashion, it seemed to induce tension in the unnamed author of the Marlborough Gallery press release. Detailing the technique through which these photographs acquired their "painterly quality by combining a minimal setting with a process of hand-applied silver emulsion on Thai Mulberry paper," the author compared Pagliuso's portraits to the "neo-documentary, straightforward style of German photographer August Sander," and then went on to explain (with a sort of textual moue) that the "inspiration for the subject matter came from the artist's father who had the unlikely hobby of raising show chickens in suburban Southern California."[11]

While the Marlborough Gallery may well have used the reference to August Sander's work as legitimation strategy for Pagliuso's cross-over aesthetic, his enigmatic Westerwald portraits of farmers and tradespeople serve as a bridge between the straightforward exhibition portraits of Harrison Weir and his ilk and the artistic chicken portraits of Jean Pagliuso. Part of the activist wing of the artistic movement called Neue Sachlichkeit, or the New Objectivity, Sander was distinguished by his "monumental aspiration to create a photographic typology of Weimar society, which he called *Citizens of the Twentieth Century*."[12] Sander's work has been the subject of much critical discussion, most notably by Walter

Benjamin in his 1931 essay "A Short History of Photography" (Benjamin 1972). Benjamin provides an acute and empathetic description of what he praised as Sander's "extraordinary corpus": "Sander starts with the peasant, the earth-bound man, and takes the reader through all the strata and occupations, up to on the one hand the highest representatives of civilization and on the other down to imbeciles'" (22). Such portraits as "Two Peasant Girls" exemplify the remarkable lucidity of Sander's images. The individuals' direct, expressionless gaze is accented with small, telling details such as the handkerchief in the pocket of the dark-haired girl and the girls' tightly clasped arms as they face their photographer. These touches undercut or, as some critics have argued, exaggerate the clinical objectivity of the camera's classifying lens.[13]

While one might explain these specific aspects of Sander's photographs by linking them to the era of photography's rural beginnings, when the need for a lengthy open-air exposure time produced a near expressionlessness in the subjects, Benjamin argues rather that these photos mark the moment of transition from the preindustrial era to the rise of industrial capitalism, and from the medium of portraiture to that of documentary photography. To Benjamin, Sander's photographs are simultaneously appealing and dangerous in their nature as virtual taxonomies. Borrowing from Goethe, Benjamin describes their technique as "a delicate form of the empirical which identifies itself so intimately with its object that it thereby becomes theory." Sander's acute observation of the physicality, class location, and professional status of his subjects appears to grant them that "peculiar web of space and time" Benjamin called the aura. Yet the information about spatial and social location is highly charged: Sander's subjects are frequently the citizens of a contested physical and socioeconomic geography: the Westerwald, that rural area of forested mountains, farming, and small villages where artisans, craftspeople, peasants, and shopkeepers lived as they had for generations.

Documenting the ways of life of such citizens, Sander captured a premodern way of life already in the process of drastic change, as craftsmanship and artisanal labor guided by local knowledge were being replaced by industrial and scientific processes relying on expertise and abstract knowledge. Perhaps that is why Benjamin observed in 1931 that the photographer was "descendant of the augurs and the haruspices."[14] Geographically, sociologically, arguably even biologically, Sander reveals aspects of human life in the process of disappearing while also foretelling modes of life to come. Sander documents citizens who are—as he understands it—on the margins of society: the unemployed, the blind, the infirm, and the congenitally disabled as well as those with marginal livelihoods such as the street sweeper or the beggar, and liminal professions in such venues as the carnival, the circus, and the theater. By placing these photographs in a sequence, along with the writer, the painter, the architect, the industrialist, soldiers, society ladies, a cleaning woman, a Protestant priest, and a Storm Trooper chief, Sander created a taxonomy of what he called the "Citizen of the Twentieth Century."

Scholars have argued all positions in the heated debate about the ideological implications of Sander's taxonomizing. Yet they all agree that the practice of creating taxonomies is a perilous one.[15] As the circumstances of the time change, Benjamin suggests, the nature of Sander's photographic gaze can shift suddenly from a transparent and even loving empiricism to instrumentality or even oppression. "Work like Sander's can assume an unsuspected actuality overnight. Shifts in power, to which we are now accustomed, make the training and sharpening of a physiognomic awareness into a vital necessity. Whether one is of the right or the left, one will have to get used to being seen in terms of one's provenance. And in turn, one will see others in this way too. Sander's work is more than a picture-book, it is an atlas of instruction" (Benjamin 1972, 22). Benjamin reiterates Doblin's assertion that Sander's life-work has a scientific dimension: "Just as there is a comparative anatomy which enables one to understand the nature and history of organs, so here the photographer has produced a comparative photography, thereby gaining a scientific standpoint which places him beyond the photographer of detail'" (ibid.). The very quality of Sander's regard—its physiognomic awareness—can serve those in power, Benjamin argues, because it can instruct the viewer on the "provenance"—the social and biological origins—of the photographic subject.

To be sure, instruction in one's biosocial location was also conveyed by the illustrated taxonomies of naturalists such as Tegetmeier and the unnamed artist(s) who illustrated the poultry breeds in Darwin's *The Varieties of Plants and Animals under Domestication*. These engraved, drawn, or painted images carry information about both space and time: the space and time of the present chicken in its biological and social niche as well as the potential space of some future specimen. Yet there is a change in scale that accompanies the change in medium. To Benjamin, photography (as opposed to earlier modes of representation) is inherently implicated on what we might call the microscopic scale: "Structural qualities, cellular tissues, which form the natural business of technology and medicine are all much more closely related to the camera than to the atmospheric landscape or the expressive portrait" (Benjamin 1972, 7). Benjamin's analysis of Sander's portraits calls to mind the work of performance theorist Richard Schechner, so let's pause there to consider the implications of the resemblance.

Pointing to recent studies by Victor Turner and others that "signal a convergence of anthropological, biological, and aesthetic theory," Schechner proposes that performance exists at seven ascending magnitudes, from the unambiguously biological to the manifestly social (2003, 322). His schema for these ascending magnitudes charts a chain of incremental transformations, from brain event, microbit, sign, scene, and drama, to macrodrama (325–326).

Schechner encourages scholars to question the variety of ways any performance can move between those magnitudes: "Is context the determining factor—so that performances actually 'build down' from larger meaningful units to smaller and smaller meaningless 'performables'? Brain events . . . are not themselves units of meaning—they are like phonemes and words that acquire meaning only used in sentences or bigger semantic units. Thus performances of small magnitudes gather meanings from their contexts. Or perhaps it is more complicated. Meanings may be generated and transformed up and down the various magnitudes" (323). This notion that the smallest, "microbit" magnitude of performance, including the "brain event: the neurological processes linking cortical to subcortical actions," can be linked up the scale to changes at the magnitude of populations suggests just how it is that photographs may convey something omitted from earlier types of breed or individual portraiture. Whether part of Sander's compilation of citizens or John H. Robinson's group portraits of chickens and of male and female breeders and judges in his poultry book, each photographic subject is presented as a documentary focus rather than as a personality. And yet each portrait of an individual contains information that can be read up or down the scale of magnitudes to tell us about the specific embodied self caught in the lens as well as the set of traits that individual shares with many others. So a sequence of these implicitly documentary photographs can enlist the eye of the selective breeder-exhibitor (in the case of the chickens) or that of the eugenicist (in the case of the exhibitors and judges). Intervening on a macro- and micro-social level, these photographs reveal how populations can be and are shaped by changes in the conception and development of a single organism.

Chicken Culture

In *Keywords*, Raymond Williams defines the word culture as: "one of the two or three most complicated words in the English language . . . mainly because it has now come to be used for important concepts in several distinct intellectual disciplines and in several distinct and incompatible systems of thought" (1983, 87). When the Marlborough Gallery compared Jean Pagliuso's "Poultry Suite" to August Sander's documentary photography, it did more than provide a legitimating photographic lineage for a show that might confuse its metropolitan audience with its extra-artistic references. It also explicitly reasserted disciplinary authority over the transgressive, interdisciplinary pull of Pagliuso's own agricultural narrative, which still seems to be perceived within the art world as what Williams terms a "distinct and incompatible [system] of thought." Exploring Pagliuso's work with an eye to the relations between art and agriculture, we have found there an expansive set of connections between the practice of breeding, particularly selective breeding, that drew on poultry portraits as its "atlas of instruction," and the successive waves of cultural representations of chickens, as objects of aesthetic and economic value, sources of regional and national pride, and evocative images of human potential. Unlike the unitary relationship between character and occupation memorialized by Sander's early-twentieth-century documentary photographs, Pagliuso's early twenty-first-century "Poultry Suite" documents an increasingly flexible set of associations between human and avian traits and practices that together enliven the notion of chicken culture.[16]

Disability

"Which came first, the chicken or the egg?"

–Anonymous

One chilly night near the end of December 2007, I sat in a cramped theater watching a chicken crowing on stage. To be more precise, I was watching a woman acting the role of a disabled man crowing like a chicken. Celeste Moratti, an Italian actress living in New York City, was playing the role of Antonio, a psychiatric hospital patient who believed he was a chicken. Chickens and disability: this unlikely combination of themes had drawn me to the La MaMa Experimental Theater in New York's East Village to see the American premiere of Italian playwright Dario D'Ambrosi's *Days of Antonio* (*I Giorni di Antonio*, 2007).[1]

Mental disability has long been the central subject of D'Ambrosi's work and of the theater group he founded, the Teatro Patologico. This group draws its inspiration from the antipsychiatry activism of Dr. Franco Basaglia, who founded an organization to reform mental health treatment in Italy based on the position that "psychiatry is politics, that psychiatry provides scientific support of the existing establishment, that scientific neutrality is a myth, and that existing standards of normality and deviance result in the oppression of certain groups in society." Beginning in 1979 with the passage of what was informally known as the Basaglia Law, this crusading physician inspired a comprehensive reform of the Italian mental health system. Hospitalization decreased drastically, and inpatients were released, reflecting the new view that "the problem was the institution itself—its authoritarianism, hierarchy, inflexibility, and medicalization of 'social' problems" (Mosher 1982, 200).

Playwright Dario D'Ambrosi responded to this dramatic new movement for psychiatric reform by signing himself in for a three-month stay as a voluntary inpatient in the Milan psychiatric hospital Paolo Pini. He hoped to write a play that would lay bare the reality of mental illness as a social, political, and cultural system, an impairment that society made into a disability. In Paolo Pini, in a patient file pilfered from a doctor's briefcase, he found the story that would become the core of *Days of Antonio*: the tale of a disabled boy born in 1916 a small town near Milan. Because he was born with one leg much shorter than the other, the boy's parents kept him exiled from their home, confined in the chicken coop, releasing him only to be exhibited—for pay—as a sideshow freak, when they wanted extra money. By his teenage years, Antonio had come to think of himself as the king of the roosters, and enjoyed *droit du seigneur* over the hens. Yet when his parents found him with a prostitute, it was that human transgression rather than his sexual encounters with the hens that led to his psychiatric incarceration.

D'Ambrosi's play begins on the day Antonio arrives in the psychiatric hospital and follows his hospitalization as he gradually realizes his outcast nature: "he's not an animal and will never be a human being" (Slaff 2007). While the play is nearly wordless, for Antonio's only means of self-expression are the squawks, pecks, and scratches of his avian delusion, even a brief transcription will reveal D'Ambrosi's vivid, nearly surrealistic evocation of the effects of institutionalized psychiatry on the experience of mental disability.[2]

The Performance

The play opens with the young, pajama-clad Antonio huddled on the floor of a sparely furnished room. He is crowing, scratching, and pecking on the wooden floorboards between two hospital beds. Another man, also in pajamas, is sweeping the floor. Gesturing as if to sweep him under the bed, his roommate, Giacomo, says, "Antonio, there are very precise rules in here and they are to be respected." Antonio answers by grunting, scratching, and peeping. Giacomo whistles in response. Surprisingly, the two are soon performing a duet, which they seal with a kiss. Yet when Giacomo begins to paw Antonio, the kiss turns ugly. Although a nurse catches the abuse as it happens, Giacomo denies it, saying it was Antonio who touched him "there." "I am the head physician's favorite," Giacomo announces. He adds, in an apparent non sequitur, that he went mad when "they" put nails into his head.

The next scene occurs in the dayroom. The nurse is dancing a tango with the doctor while they discuss Antonio's treatment. "Antonio is around eighteen years old," the nurse explains. "He was admitted because he had raped a woman. One leg ends where his knee should be. He has always lived with the chickens." The hospital staff's indifference to Antonio contrasts with the kindness of his roommate Giacomo, who offers him water in a tin cup. When Antonio's beak makes him unable to drink it, Giacomo unrolls a scroll of chicken wire and leaves the stage.

Suddenly we are in Antonio's past. Antonio's mother comes onstage to the accompaniment of lush romantic music. Wearing a long brown skirt and shawl and carrying a basket of laundry, she scatters corn for the hens. "Don't play with the chickens," she admonishes Antonio, and muses, "God gave us this misfortune and there's nothing we could do to change it. Maybe if the corn harvest is good, we can put another room on the house and you can come inside. But you seem to like it out here with those chickens." Just then, two men appear carrying a sign reading "Marchetti Brothers." "He looks just like a rooster!" one of them exclaims. "I bet he can crow just like a rooster!" With that, they pay Antonio's mother for the first of Antonio's sideshow performances.

The scene changes again. We are back on the hospital ward in the middle of the night; lights are low, and the roommates are asleep in their beds. A man in a white jacket enters with a nurse, saying: "It has been one month since Antonio was brought to the Asylum. They've found out he didn't rape a woman. A prostitute took advantage of him. The parents have refused to take him back." Then the white-jacketed man pokes Giacomo's forehead with needles, as if to lobotomize him, and we realize that Giacomo's seemingly insane accusation that "they" put needles in his head is accurate.

Although still inmates of the psychiatric hospital, Antonio and Giacomo carve out lives for themselves, negotiating the boundaries of human and animal, sanity and insanity. We see them in the next scene as the dawn comes. Antonio crows, then Giacomo crows too. Giacomo washes Antonio's hands, face, and ears and offers him lessons in talking, standing, and even walking, with a walker he has build out of a broken hospital chair. "If you can stand, I'm going to take you out of here. You can go out!" he urges. Such an escape is impossible for himself, Giacomo confides, because he has been lobotomized. The nails inserted in his head by the staff have caused internal voices, all saying "NO!"[3] Giacomo gives Antonio a bright yellow plastic chicken. Antonio, brought to tears by the gift, sets the chicken on the floor between them. "I love you," confides Giacomo. They watch together as the chicken lays an egg, to the accompaniment of a cheerful popular tune.

In the play's enigmatic final scene, a significant moment of recognition is cut short by the unmasking of a charade. It is nighttime on the ward. Antonio wakes. Using the walker Giacomo has built for him, he goes to the mirror where he stands, staring at his reflection. The white-coated man returns, looks at Antonio gazing at himself in the mirror, surveys the plastic chicken and its egg, refers to the papers in his hand and announces: "Today, July 2, 1920, at the age of eighteen, suffering from pulmonary emphysema, Antonio dies." Just then, the nurse enters. She grabs the case notes from which the man was reading and tears them into little pieces. "Your game is over," she scolds him. "Go back to the pavilion with the other patients." Unmasked, the "doctor" hands his clipboard to someone in the first row and his white coat to another member of the audience, and walks offstage. As the song of an Italian tenor fills the room, a single yellow spotlight illuminates the chicken and her egg, alone on the stage.

Representing Disability

Disability is an issue that merits serious discussion, but—as E. B. White found—some New Yorkers "regard the hen as a comic prop straight out of vaudeville" (White 1944, v). After the curtain fell on the play's New York City premiere, I found myself musing on D'Ambrosi's decision to give the chicken such a central place in his fierce critique of psychiatry while I savored the play's brush with vaudeville and the sideshow as well as its challenge to the unstable boundaries of species, sexes, and abilities. Chickens aside, just to represent disabled people on stage appropriately is a difficult proposition. To tell Antonio's story, D'Ambrosi would have had to deal with a number of issues. Could he "narrate the disabled body without replicating the static model of a diversity of biologies"? Could he "avoid recasting disabled experience as another landscape of voyeurism, equivalent to earlier freak show spectacles?" Most poignantly of all, how could the play tell its story of disability "without recourse to the pathologizing discourse of interventionist medicine, on the one hand, or to the grotesque, on the other?" (Snyder and Mitchell 2001, 381).

In his delusion that he is a chicken, as in his physical disability, Antonio is not *normate*. Disability studies scholar Rosemarie Garland Thomson explains that "this neologism names the veiled subject position of cultural self, the figure outlined by the array of deviant others whose marked bodies shore up the normate's boundaries." As "the social figure through which people can represent themselves as definitive human beings," the normate operates by exclusion (Thomson 1997, 8). Antonio's confinement in a mental hospital arguably demonstrates the process by which modern medicine *produces the normate*, by turning anomaly into disease (Kilpatrick 2004–2005).[4] In an illuminating survey of the field of disability studies, Sharon L. Snyder and David T. Mitchell trace the development from analysis of the production of a normative body through the harnessing of new medical specialties to the exploration of the transgressive, excessive, and liberating potential of an unclassifiable nonnormate body. With the rise of modern medicine, they argue, different bodies came to be seen not as simply anomalous, naturally occurring variations, but as pathological and deviant entities that must be categorized and prevented, if not cured. First body studies and then disability studies scholarship were concerned with documenting the ways "the articulation of bodily differences shifted from entries in a catalogue of biological diversity—a medieval approach to the study of Nature—to an objectifying taxonomy of deviance" (Snyder and Mitchell 2001, 371).

From a disability studies perspective, then, D'Ambrosi's very staging of disability as well as the play's passing allusion to circus sideshows risks inviting the stares of the curious and even positioning Antonio as a sideshow freak. At one time, stares and enfreakment were the norm (Thomson, 1996,

53–80). Yet over time, representations of people with disabilities have changed, reflecting the ascendance of different explanatory models for disability, as Paul Longmore has detailed. In the moral model, dominant through the Middle Ages, deformity and impairment were seen as the result of a religious or spiritual failure—an individual or collective sinful thought or action. Then in the Renaissance the medical model emerged, which viewed disability as an individual's illness, which it was a doctor's duty to alleviate. Building on the growth of orthopedics in the late nineteenth century, that model required medicine (both private and state) to treat and, if possible, cure people with impairments, who were understood to be defective (Kowalsky 2005). Similarly, in the United States, people with disabilities were subject to immigration barriers, eugenic sterilizations, and forced institutionalizations, all following from the same medical understanding of their defective nature (Stiker 1999; Baynton 2001). In the 1960s, with the emergence of the disability rights movement focused on self-determination and civil rights, and the subsequent development of the disability studies movement that reframed disability as a "social phenomenon, social construct, metaphor and culture," a shift occurred away from the medical model with its emphasis on the "prevention/treatment, remediation paradigm" toward the "social/cultural/political paradigm." This new paradigm conceptualized impairment as a state that is manifest physically or cognitively/emotionally, and it understood disability as the result of a societal failure to accommodate any kind of impairment.[5]

While the social model was enormously productive, generating not only the activist accomplishments of the disability rights movement but the intellectual contributions of the academic field of disability studies as well, its rejection of the medical model as individualizing and pathologizing has recently been challenged by a number of disability studies scholars. They argue that the social model alone may not be adequate for representing the experience of people with disabilities. Because it tends to shy away from discussions of the experiences of medicalized disabled bodies, they argue, it results in a curiously abstract perspective from which the day-to-day texture of life with an impairment is missing (Bérubé, n.d.; Siebers 2008). Sociologist Bill Hughes, who has argued forcefully in his earlier work against the geneticization and aesthetic invalidation of disabled people, in a 2002 essay held that the social model ignores the significant role culture plays in shaping bodily meanings. As he sees it, the disability studies field has taken a wrong turn in its overwhelming adoption of the social model because it has constructed a new dualism opposing impairment, which is viewed as exclusively presocial, to disability, which is viewed as exclusively social. The result is, he argues, a "reluctance to articulate the social constitution of impairment," which leads to an unfortunate syllogism: "If impairment was the opposite of disability, and disability was socially constituted, then impairment must be biologically constituted" (Hughes 2002, 63).[6] The effect has ironically been to remedicalize much of the experience of disability, Hughes argues: "Impairment must, therefore, be taken to refer to that palpable and pathological fleshly object that constitutes the subject matter of medical science. . . . Thus the social model, conceived of as the intractable opponent of all things associated with the medical model of disability, came to share with it a common conception of the body as a domain of corporeality untouched by culture" (67).

Given the challenges inherent in representing people with disabilities without pathologizing them or rendering them as grotesque, D'Ambrosi arguably made things even more difficult for himself by choosing a protagonist who combined physical and mental disabilities. Mental disability has historically been a difficult topic for disability studies, with few scholars giving it a primary focus. As Longmore has observed, "The framing of 'disability' by some [disability studies] scholars in terms that fit some 'disability' experiences, namely phiz diz [physical disability], but not others, has been criticized. . . . These attitudes and actions evidence the reality of hierarchies of disability."[7] Academic scholars in particular may consciously or unconsciously subscribe to such a hierarchy of disability since their intellectual investment in the life of the mind may result in a preference for the study of physical disabilities over that of mental and cognitive disabilities.[8] And even within the realm of mental disability, hierarchies persist. Examining filmic representations of mental illness in relation to the times in which these films and their advertisements circulated, Elizabeth Donaldson has demonstrated that the diagnostic psychiatric gaze has produced a hierarchy even within mental disability: certain modes of mental illness are increasingly normalized (depression) while other forms are maintained as abjectly abnormal (schizophrenia) (Donaldson 2005). The different social spaces allotted to different forms of mental illness are captured even in a publicity photograph for D'Ambrosi's play: the blankly staring lobotomized Giacomo and the smiling white-coated "doctor" ring a distraught, bruised, agonized Antonio, encased in chicken wire and surmounted by the little yellow plastic chicken.

As I walked home from the theater that night, I mused on D'Ambrosi's decision to take on both aspects of Antonio's disability: physical (as his deformed leg exiled him from human society to the family poultry yard) and mental (as he experienced the impersonal regimentation and neglect of institutionalized psychiatry). I understood how the theme of chickens linked both kinds of disability as the play moved from the chicken yard of Antonio's youth to his adolescent days in the mental hospital, where Giacomo's precious gift of the yellow plastic hen provided a welcome affirmation of his personal mental world as well as a counter to the hospital regimen of impersonal abuse. I began

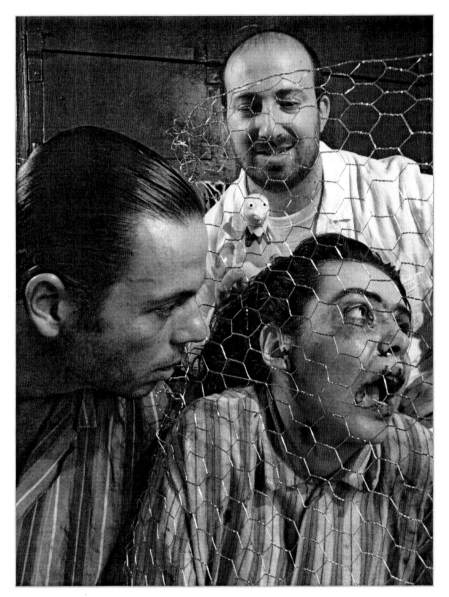

FIG. 13. *Days of Antonio* publicity photograph.

Photo: Jonathan Slaff

to grasp how *Days of Antonio* used chickens to raise questions of origin, identity, and community. But I found myself still mulling over the play's last image: the chicken and egg, illuminated by a yellow spotlight, on the bare stage. Was D'Ambrosi posing his own version of the ancient riddle, "Which came first, the chicken or the egg?"

The Chicken and the Egg: Embryology, Teratology, Eugenics

The riddle "Which came first, the chicken or the egg?" dates back to "Plutarch's *Symposiaques, or Table-Questions*," whose first question is "Whether was before, the hen or egg?" In his wide-ranging *History of Embryology*, Joseph Needham details the central role played by hen's eggs in the development of the field (Needham 1959, 68). The unknown author of the collection of Hippocratic writings, whom Needham described as "the first embryologist" (31), advocated a series of experiments on hen's eggs as a way of understanding human development, in particular the relation between our early development and our later status as human beings, and our relations as human beings to other animals: "Take 20 eggs or more and give them to 2 or 3 hens to incubate, then each day from the second onwards till the time of hatching, take out an egg, break it, and examine it. You will find everything as I saw in so far as a bird can resemble a man. He who has not made these observations before will be amazed to find an umbilicus in a bird's egg" (36).

Aristotle, who systematized embryology, also opened hen's eggs at different intervals to discover the sequence by which the organs develop. His *On the Generation of Animals*, "the first great compendium of embryology ever produced," according to Needham (39), compiled so many insightful observations about the embryo that it influenced "the next twenty centuries" of scientific research (54). With the work of William Harvey in the seventeenth century, the field moved from a static, descriptive one, preoccupied with "the study of the embryo as a changing succession of shapes," to a dynamic, experimental one focused on "the study of [the embryo] as a causally governed organization of an initial physical complexity" (116–117). So, in his *De Generatione Animalium*, Harvey explained, "We have already discovered the Formation, and Generation of the Egge; it remains that we now deliver our Observations, concerning the Procreation of the Chicken out of the Egge. An undertaking equally difficult, usefull, and pleasant as the former" (quoted in Needham 1959, 47, 117). From Harvey's embryological experiments, a lineage extends to modern and contemporary developmental biology and genetics discussed in the previous chapter—both C. H. Waddington's work on chick embryos in the 1930s and the more theoretical speculations of Weiss and Buchanan.

Rather than following that lineage now, however, I want to pick up a less-known route. The final chicken and egg image in *Days of Antonio* also brings to mind another field that relied on the same model organism: teratology, or "the production of human monstrosities" (Cooper 2004, 12). While embryology is usually defined as focusing on normal development of the embryo, teratology studies the embryo's abnormal development. Despite this theoretical division, however, in practice the fields have significant overlap since "whatever an experimental investigator does to an embryo (and in many cases the experimentalist may be Nature) some things go right and some go wrong, [thus] all distinctions between experimental embryology and experimental teratology become blurred; the disciplines are symbiotic" (Oppenheimer 1968, 146).

Descriptive and experimental teratology originates with the transcendental anatomist Étienne Geoffroy Saint-Hilaire, whose interest in experimentation on chick embryos dated from an expedition he made to Egypt in 1798 with Napoleon Bonaparte (Richards 1994).[9] Geoffroy joined two other scientists at the newly established Institute of Arts and Sciences in Cairo, where he hoped to carry out experiments on the problem of sex determination (Oppenheimer 1968, 148). To Geoffroy, chickens seemed an ideal model organism: they were easily available and produced a generous supply of eggs that could be incubated artificially after manipulations to test the theory that "long eggs produce male chicks, short eggs females. If one could change the shape of the egg, one could change the sex of the bird" (149). And with a tradition of artificial incubation of hen's eggs in clay ovens that dates back to 1400 BC, Egypt offered a ready pool of expert technicians and suitable incubators, ensuring that manipulated embryos could be held successfully under artificial heat until hatching (149). Geoffroy had grand plans for extensive experiments on sex determination and requested from the Institut d'Égypte (the scientific institute he had been brought to Egypt to help establish) "an incubator, 500 to 600 chicken eggs, an enclosure for 500 to 600 chickens . . . measuring instruments, and two Egyptian assistants" (Appel 1987, 76–77).

Geoffroy never actually carried out these experiments; personal health problems led him to return to France instead. But in later years his experimental interests expanded to include "the causes of malformations and the laws governing their production" and he was launched on a road that would place the avian egg at the center of this field so foundational to the medical approach to impairment and disability (Oppenheimer 1968, 150). He was particularly engaged with the widespread debate at the time between preformation and epigenesis—two theories explaining development as either predetermined in the egg from the beginning or unfolding without a preordained direction. Convinced that arrested development produced the abnormalities that were his focus, beginning in 1820, he began to carry out experiments on whole hen's eggs. He covered the eggs in goldbeater's skin, the transparent outside membrane of an ox intestine having the tensile strength to hold the egg together, and then shook them, changed the temperature at which they were incubating, or even incubated them with small openings cut in the eggshell (150). Since his time in Egypt, Geoffroy had been drawn to speculative and grand morphological theories (151). He believed that not only had he caused monsters to be produced in ordinary hen's eggs but that he could now specify precisely which kind of malformation would result from any particular intervention in the incubating egg. "Geoffroy argued by analogy from the production of monsters to the

origin of species. . . . He was . . . convinced that he had experimentally demonstrated the way new species arose in nature" (Richards 1994, 381–382). While
the resulting embryos did demonstrate some developmental anomalies, the
experiments were less than completely conclusive. As Toby A. Appel observes,
Geoffroy "had as yet only an intuitive notion of an underlying unity of plan, no
real explanation for similarities of structure, and no method for making rigorous comparisons" (Appel 1987, 81). This tradition of experimentation on hen's
eggs was maintained by Geoffroy's son, Isidore Geoffroy Saint-Hilaire, who published a *Treatise on Teratology* (*Traité de Tératologie*) in 1832. With that publication, Isidore gave this new field a distinct identity, naming it "teratology" and
thus distinguishing it from embryology by emphasizing its focus on the production and systematic description of monstrosities.

Where Geoffroy father and son left off, embryologist Camille Dareste picked
up, also working on hen's eggs but carrying out experiments that were more
extensive and more systematic. Covering the eggs with varnish, oil, or wax;
incubating them in chambers that heated the sides to different temperatures;
even shaking the eggs in a candy-making machine created to pack filling into
chocolates, Dareste attempted to induce abnormal development, and the
results of these experiments convinced him that higher than normal temperatures were a highly effective way to produce monstrosities (Oppenheimer 1968,
155). With the publication in 1871 of his *Recherches sur la Production Artificielle des
Monstruosités; ou, Essais de Tératogénie Expérimentale*, Dareste provided a set of
first principles for this field that he hoped would be a "science of all possible
bodies" revealing an "unlimited variability" of forms (Cooper 2004, 200). As he
explained, his very first problem as a teratologist was the creation of his
research subjects, "in other words, the artificial production of anomalies and
monstrosities" (Dareste 1891, 7; my translation). Indeed, in founding the field of
teratology, Dareste invented the new "embryology of monsters" (Oppenheimer
1968, 126).

Like Geoffroy *fils* before him, Dareste was explicit in his preference for the
chicken's egg, not only because it was easily available and external, unlike most
other ova, but also because it could be incubated artificially. The application of
his experiments was of less interest to him; he left to future scientists the problem of deciding whether his avian embryology illuminated the development of
other species. "Not to prolong my research indefinitely, I limited myself to one
species only. Hen's eggs, which can be obtained in as great a quantity as one
wishes, lend themselves easily to the study of normal and abnormal development. . . . Time will tell whether the results I have obtained by studying one single species can really be applied to the whole branch of vertebrate mammals"
(Dareste 1891, vii–viii; my translation).[10]

While Dareste's teratogenic experiments all focused on whole hen's eggs,
later researchers intervened in various ways in embryonic development directly

within the egg, anticipating techniques used by modern embryologists, and later by fetal surgeons. Working on the egg of the domestic hen, Stanislas Warynski and Hermann Fol hand-drilled 2–3 cm diameter holes in the shell, working first with dental drills, and later with a sharp scalpel. Placing the egg on its side, so the embryo lay on the part of the shell that had not been cut, they kept the eggs in incubation, turning them regularly. Then they removed them from the incubator for fifteen minutes at a time and performed "direct surgical interventions" within the shell in the attempt to produce a number of specific deformations in the developing embryo (Oppenheimer 1968, 156).[11]

Dareste's embryological experiments convinced him to view anomalous development not merely as just another variation on life but as "a time arrow of innovation . . . [to conceive] of evolution as an expanding horizon of variability, an open ended exploration of the possibilities of life" (M. Cooper, 16). Clearly monsters were more than merely evidence of nature's endless variety, however; they became "harbingers of evolutionary change" or, as he termed them in his papers, "êtres ébauchés, preparatory or precursory beings" (Richards 1994, 381). Nineteenth-century teratogeny, like the larger field of teratology within which it was nestled, viewed its role as exploring the unending possibilities of unfolding life. Yet with the discovery of Mendelian genetics, teratology's fascination with infinite variation was succeeded by a new focus on the inherited genetic disorders.[12] Now, although "the same inherited disorders studied by the eugenicists continue to occupy modern clinical geneticists," anomalous development came to seem less a fascinating question of developmental laws (calling for a systematic survey of variations) than an urgent question of medical responses.[13] Monstrosity (or developmental abnormality) offered a key to normal development.

While the medical approach to developmental malformations predominated in the twentieth century, the descriptive origins of teratology as a field remained even as the anomalous was being rethought as the pathological. Teratology as a field did not disappear; it merely reconfigured itself. Replacing delight in organic excess with a normalizing therapeutic agenda, contemporary teratology emphasizes the detection and prevention of congenital anomalies. By 1967, prime among its concerns were the effects of modernity itself on the human body: the effects of "radiation, viruses, and chemical substances" as well as "the implications of vitamin deficiency and hormone metabolism." Avoiding congenital disability became a crucial part of teratology's mission, as described in the conference proceedings of a 1967 symposium: "The intention of the Italian Society of Experimental Teratology has been to focus attention on the various problems of embryology and teratology and to establish sufficient discriminatory criteria to distinguish those factors which cause unjustified general alarm, from those which threaten real danger to the life of the unborn child" (Bertelli and Donati 1969, vii). The symposium drew together researchers in disciplines ranging from plastic and reconstructive surgery, pathology, and embryology to zoology, genetics, and pharmacology, all focused on "the solution of problems related to the origin and prevention of pathological deviations of somatic development" (vii). By 2005, teratologist Richard Jelenk advocated that contemporary teratology return to, and adapt, Camille Dareste's experimental principles, which he argued could remedy contemporary teratology's failure to record any "triumphant success even when applying powerful techniques of molecular biology and genetics."[14]

Eugenics and the Triumph of the Egg

I have argued that significant connections exist between Dareste's teratological experiments on avian eggs in the nineteenth century; the work of Alexis Carrel, Peyton Rous, and Montrose Burrows on avian sarcomas in the early twentieth century (touched on in chapter 2); and this early twenty-first-century notion of harnessing the proliferative, potentially innovative form of tissue growth found in teratomas for regenerative medicine. Before we accept that as the comprehensive story of the role of chickens and eggs in the construction of disability, however, we need to return to another era important to the development of the medical treatment of disability: the heyday of eugenics, the (pseudo)scientific discipline dedicated to the improvement of the human population through means of negative and positive interventions. Eugenics has been the subject of extensive and illuminating analysis, especially by historian Daniel Kevles, and I will not attempt to address its full significance for disability studies. Eugenic regulations both restricted immigration of foreigners with disabilities and constrained the life options of Americans with disabilities, even as the regulations were used to boost the opportunities of other oppressed minorities (Baynton 2001).

When the model for D'Ambrosi's Antonio was born in 1916, chickens were playing another role in the investigation of inherited illness and disability: they anchored two of the foremost sites of eugenic research in the United States, the Station for Experimental Eugenics and the related Eugenics Record Office, at Cold Spring Harbor, New York. Two of the leading American eugenicists were drawn together by a shared interest in chicken breeding. In 1907 Harry H. Laughlin, the superintendent of public schools in Kirksville, Missouri, wrote Charles Davenport to report that he was "making some interesting experiments by crossing White Fluffs or Klondykes and the Long Tailed Yokohama fouls [sic]." Laughlin explained that he had succeeded in breeding "several birds that to all appearances are Yokohamas save that they are pure white in color. . . . Poultry breeders can give me now no information and so I wrote asking whether or not the Phoenix Indian Game, the Shomo or Yokohama, and the Tosa fouls [sic] are one and the same breed" (Laughlin 1907). The search for pure white chickens that drew Laughlin to Davenport was more than simply a preference for a specific color in avian plumage: it was an aspiration to improve the human species. Although eugenicists "can't mate men and women as we please, like cocks and hens," as Francis Galton, the founder of the Eugenics Education Society, wrote William Bateson in 1904, "we could I think gradually evolve some plan by which there would be a steady, though slow amelioration of the human breed" (Galton 1904). Chickens were a central part of the Cold Spring Harbor mission. Only a year after Galton's letter to Bateson, Davenport presented a paper, "Inheritance in Poultry," at the American Breeder's Association, and within a year of Laughlin's introductory letter to Davenport, chicken houses had become a prominent feature on the lawn behind the Italianate main building of the Cold Spring Harbor Research Station.[15]

With the encouragement of the American Breeders' Association, researchers eager to move beyond theoretical speculations about hereditary diseases began to return to the strategies of their teratological ancestors, finding in poultry research a first-rate opportunity for scientists who wanted to study how defects could be deliberately created through controlled breeding. So eugenicist Lucien Howe advocated poultry research as a way to discover the causes of hereditary blindness, in an article in the *Journal of the American Medical Association* published in 1918. "Theoretical knowledge is . . . second hand mental furniture. The best way to learn these principles and one vastly more interesting is to supplement the reading with at least a few experiments. *The breeding of eye defects is easier than most persons imagine.* Chickens and pigeons are the best subjects for such experiments. By advertising in the Reliable Poultry Journal and other trade papers, it has been possible to obtain for the parent stock more than a dozen specimens of eye defects. . . . The different pens of chickens at . . . Mendel Farm . . . have proved to me a source of much interest and enlightenment" (Howe 1918, 1996).

The same year that Howe published his article advocating poultry research as a key to birth defects, Sherwood Anderson published his classic short story, "The Triumph of the Egg."[16] Published in *The Dial* only two years later, this tale in which a hen's egg serves as embodiment of both the sublime potential of life and its limitations offers a first-person narrative of a young boy's struggle with identity, sexuality, and the passage to adulthood via a series of encounters with chickens and eggs (Anderson 1920). The boy's father is a farmhand turned chicken farmer who has become obsessed with the "little monstrous things that had been born on [his] chicken farm" (7). Although "intended by nature to be a cheerful, kindly man," as he has struggled to make a living for his family in Depression-era America the father has become fascinated with deformity and monstrosity (4). His doom-ridden temperament seems to follow from his occupation, the narrator muses: "One unversed in such matters can have no notion of the many and tragic things that can happen to a chicken. . . . A few hens and now and then a rooster, intended to serve God's mysterious ends, struggle through to maturity. The hens lay eggs out of which come other chickens and the dreadful cycle is thus made complete. It is all unbelievably complex. Most philosophers must have been raised on chicken farms. One hopes for so much from a chicken and is so dreadfully disillusioned" (5). A philosopher along the lines of natural philosopher Geoffroy Saint-Hilaire, his greatest treasure has become the collection of bottles containing the deformed chicks, "preserved in alcohol," that he dreams of exhibiting at country fairs:

> Grotesques are born out of eggs as out of people. The thing does not often occur—perhaps one in a thousand births. A chicken is, you see, born that has four legs, two pairs of wings, two heads or what not. The things do not live. They go quickly back to the hands of their maker that has for a moment trembled. The fact that the poor little things could not live was one of the tragedies of life to father. He had some sort of notion that if he could but bring into henhood or roosterhood a five-legged hen or a two-headed rooster his fortune would be made. He dreamed of taking the wonder about to county fairs and of growing rich by exhibiting it to other farm-hands.   (7)

When the chicken farm fails, the collection of monsters takes pride of place in the restaurant his father buys, just opposite the railroad station that, they hope, promises a good source of diners. "All during our days as keepers of a restaurant . . . the grotesques in their little glass bottles sat on a shelf back of the counter. Mother sometimes protested, but father was a rock on the subject of his treasure. People, he said, liked to look at strange and wonderful things" (7).

In prose mingling echoes of the Bible with the choppy diction of an advertising salesman or a sideshow barker, Anderson's narrator recalls how his father accepts the new gospel of positive thinking. He adopts "a cheerful outlook on life," and becomes obsessed with the idea that "both he and mother should try to entertain the people who came to eat at our restaurant." The "spirit of the showman" seems to be lurking somewhere in him, but in his son's eyes he is blighted, even feminized, by his continual preoccupation with teratology (9). As in medieval days it was thought that a woman who gave birth to a deformed child had been subjected to unwholesome influence while pregnant, so to the son there seems to be "something pre-natal about the way eggs kept themselves connected with the development of [his father's] idea. At any rate," he tells us, "an egg ruined his new impulse in life" (9–10).

His father's ruination happens on the day "young Joe Kane, son of a merchant of Bidwell" comes into town "to meet his father, who was expected on the ten o'clock evening train from the south." The train is three hours late, and Joe comes into the restaurant "to sit loafing about and to wait for his arrival." To Joe Kane, "the restaurant keeper [seemed] apparently disturbed by his presence." Yet the narrator knows he is "no doubt suffering from an attack of stage fright"; his father is determined to entertain this customer (10). So he declares that he can make an egg stand on end by rolling it to and fro in his hands. He scoffs at Christopher Columbus as a "cheat" who was unable to accomplish the same feat; "he talked, he did, and then he went on and broke the end of the egg." Mumbling "words regarding the effect to be produced on an egg by the electricity that comes out of the human body," he brags that he can create a new "center of gravity" in the egg before him. "I have handled thousands of eggs," father said. "No one knows more about eggs than I do." Proclaiming "the wonders of electricity and the laws of gravity," the father tries repeatedly to perform his miracle (11). Yet when he finally succeeds, it is too late: Joe Kane has lost interest and fails to notice. Desperate to win back the young man's interest, the father abandons the perspective of modern science, turning instead to an earlier, more mysterious world: "Afire with his showman's passion and at the same time a good deal disconcerted by the failure of his first effort, father now took the bottles containing the poultry monstrosities down from their place on the shelf and began to show them to his visitor. "How would you like to have seven legs and two heads like this fellow?" he asked, exhibiting the most remarkable of his creations (12).

Younger, the son of a businessman, and probably more urbane and pragmatic, Joe Kane doesn't share the father's fascination with monstrosity. Instead he is "made a little ill by the sight of the body of the terribly deformed bird floating in the alcohol in the bottle and [gets] up to go." Making one last try to impress him, the father proposes to put an egg "through the neck of a bottle without breaking its shell. When the egg is inside the bottle it will resume its normal shape and the shell will become hard again. . . . People will want to know how you got the egg in the bottle. Don't tell them. Keep them guessing. That is the way to have fun with this trick" (12). The father's mood seems more desperate than amused as he worries the egg roughly, sweating and swearing. Still Joe Kane decides that although he may be "mildly insane" he is probably harmless. He continues watching, and the father's trick is "about to be consummated," when a cascade of events aborts the trick at the last moment. The delayed train arrives in the station, Joe Kane leaves to greet his returning father the business traveler, and the egg breaks in the restaurateur's hand. "When the contents spurted over his clothes, Joe Kane, who had stopped at the door, turned and laughed" (13). As the charged language reveals, the father's humiliation is at once sexual and oedipal. "On the highest level, the father's final futile attack upon the invincible egg is an onslaught upon the process of life, a crazed attempt to pickle the ovum and ram it back up the uterus" (West 1968, 692).

With the staining spurt of the egg and his customer's laugh, the father is vanquished. It is the triumph of the egg, the son realizes when he wakes the following morning, that the age-old question remains unanswered: "I awoke at dawn and for a long time looked at the egg that lay on the table. I wondered why eggs had to be and why from the egg came the hen who again laid the egg. The question got into my blood. It has stayed there I imagine, because I am the son of my father. At any rate, the problem remains unsolved in my mind. And that, I conclude, is but another evidence of the complete and final triumph of the egg—at least as far as my family is concerned" (Anderson 1920, 13). In Anderson's short story, the wondrous developmental possibilities explored by nineteenth-century teratology are dispelled by the pragmatic realism of modern commercial society. The search for monstrous wonders is revealed to be a sham, and the lure of the anomalous is recontained (in the eggs, the chicks, the father, and the son) by successfully subordinating it to the culture of scientific progress represented by young Joe Kane.

Freaks and the Two Teratologies

The showman's passion and obsession with "poultry monstrosities" in Sherwood Anderson's "The Triumph of the Egg" was reworked twelve years later in Tod Browning's cult classic *Freaks* (1932). While D'Ambrosi's Antonio is freed from his parents' chicken yard when he is rented out to the Marchetti brothers' traveling circus act, and Andersons' youthful narrator is liberated from his father's teratological obsession when a sideshow trick goes wrong, in *Freaks* a traveling circus is the setting for an encounter between the worlds of the normal and the anomalous in which teratology holds the capacity to eradicate the latter to protect the former.[17] Adapted from Tod Robbins's short story "Spurs," *Freaks* tells the story of the trapeze artist Cleopatra, who is turned into a monstrous "chicken woman" in revenge for transgressing the "code of the freaks."[18] Cleopatra marries Hans, a little person, in a splendid ceremony in which the circus freaks welcome her as "one of us." But her marriage to Hans is anything but a declaration of kinship; instead, she plans to murder Hans to gain his inherited fortune. The circus freaks discover her plot, and in a horrific nighttime crawl through mud and rain they converge on her, knives in hand, and take their revenge.

Freaks has been the subject of some debate in the disability studies community over the nature of its treatment of the disabled. MGM advertised the film as "'a mystery drama [set] behind the scenes in a sideshow with strange and grotesque freaks and monstrosities playing principle roles." Yet it exceeded even the other horror films of its era by the very "plurality of 'Others'" coded as aberrant if not outright abhorrent" (Norden and Cahill 1998, 88). While the film seemed to actively invite the enfreakment of its actors, it also grew out of a conversation Browning had with Harry Earles, a short-statured actor with whom Browning had worked earlier. Earles, who suggested the plot to Browning, took the role of Hans while Browning cast other actors with mental and physical disabilities for many of the other freak show performers.[19] If the film intended to embrace the perspective of people with disabilities, it frequently failed to do so. Instead, its concluding scenes endorsed the notion that disability would be eradicated through the power of modern medicine, returning to the theme expressed in the film's opening epigraph: "In ancient times, anything that deviated from the normal, was considered an omen of ill luck or representative of evil" (Gonzalez 2003).

Freaks has an ambiguity, even a vacillation, that results from its commitment to the goals of modern, genetically informed teratology, as distinct from the nineteenth-century teratology of Camille Dareste. "Beginning with the film's opening epigraph, which simultaneously argues that 'the disabled' are the same as 'us' (the normatively positioned viewer) while promising the ultimate eradication of disability by modern teratology, Freaks labors with contradictory impulses—both to normalize and exoticise its disabled acting ensemble" (Snyder and Mitchell 2001, 380). While the father in Sherwood Anderson's "Triumph of the Egg" maintains his collection of monstrous chickens out of a sense of wonder at the infinity of natural forms, Freaks explicitly solicits the viewer's horror, especially in the sequence leading up to Cleopatra's final transformation into a monstrous chicken-woman: "The other disabled performers . . . do not hesitate to exact a cold-blooded revenge. In a nightmarish sequence replete with thunder and lightning, the title characters literally slither through the mud before slaying Hercules and mutilating Cleopatra to such an extent that she becomes a 'freak' too: a grotesque 'chicken woman'" (Norden and Cahill 1998, 88). In the film's opening, a sideshow barker explicitly challenges the conventional view of disabled people as Other: "We told you we had living, breathing monstrosities. You laughed at them, yet but for the accident of birth, you might be even as they are!" Yet by the conclusion of Freaks, the same carnival barker disavows the epistemological power of teratology even as he exhibits his monstrous attraction: "How she got this way will never be known. Believe it or not, there she is!" The camera pans down and we finally see Cleopatra, "looking like an oversized fowl, butchered beyond recognition" (88).[20] In the muddled vision of the film, Cleopatra's transformation into the unthinkably monstrous chicken can be read not merely as the freaks' revenge but also as their furious protest at a tradition of teratological experiments on hen's eggs to investigate the origins, production, and varieties of physical deformities being restaged as medical research dedicated to the prevention of disability.

While Freaks gave potent voice to the medical mission to eradicate deformity, it also testified to the phenomenon Ludwik Fleck described with such insight: how old ways of understanding the world persist even in those new ways of thinking by which they seem to have been banished (Fleck 1979,

98–100). In that sense, Browning's film was both elegiac and forward-looking in its imaginative vision. As we have seen, late twentieth-century teratologists did not surrender their interest in the production of monsters, nor had they diverged greatly from their major experimental focus on chick embryos. They were simply more attentive than Camille Dareste to the pressure for medical applications for their knowledge, not least because of the growth of the pharmaceutical industry and the wider awareness of the impact of environmental toxins since the 1930s.[21]

Such tensions are evident if we return to the proceedings of the symposium held in 1967 by the Italian Society of Experimental Teratology. In an article that recalls the work of C. H. Waddington on chick embryos, anatomist J. Mackenzie confirms that "[for] many years now, growing the chick embryo *in vitro* has been an extremely useful technique for the embryologist . . . and I am sure it has a part to play in the examination of pharmaceutical preparations for teratogenic activity" (McKenzie 1967, 43). Ruth Bellairs, of the department of anatomy and embryology at University College London—the same department where I spoke with Dr. Carlo Stern about the work of C. H. Waddington—also affirms the importance of basic research on chick embryos, concluding, "the more we can discover about the way in which these cells communicate with one another, the nearer we are to understanding how it comes about that sometimes a normal embryo develops and sometimes a monster" (Bellairs 1967, 172). Yet from the perspective of medicine's commitment to the biological norm, the stakes are clear. As A. K. Palmer puts it in his study "The Relationship between Screening Tests for Drug Safety and Other Teratological Investigations," "The advent of the thalidomide disaster added a new dimension to the . . . multi-disciplined science of teratology, namely, the development of sound laboratory methods by which similar tragedies could be prevented" (Palmer 1967, 55).

This new version of teratology may even bring disability studies scholars back to medicine. For despite its normalizing uses, it also holds the potential to teach biomedical researchers to think beyond their attachment to normalcy. Teratological research anticipated the rise of stem cell research, Melinda Cooper has argued, replacing medicine's former focus on regulating and preserving the limits on growth with a new commitment to testing those limits. She suggests that medicine is returning to the earlier perspectives of teratology and teratogeny as a way to explore the infinite possibilities in the pluripotent stem cell: "What is exceptional about recent develops in stem cell research is the fact that . . . monstrous possibilities are being exploited *as a source of regenerative tissue*. . . . In the process, the deregulated growth of the monstrosity, that ultimate counter value to normative theories of organic life, comes to represent the most extreme potentiality of life itself" (Cooper 2004, 21).

Let's pause to consider the implications of this notion: monstrosity (or, to use the contemporary term, congenital disability) is being understood not as a throwback or a diseased departure from the norm but as a potential direction for future evolution—as a harbinger of future beings. There may be broad ramifications to this new perspective on life. Evelleen Richards argues that Geoffroy's vision of "hopeful monsters" harmonized with his radical political philosophy supporting "democratic politics . . . and doctrines of self-development" (Richards 1994, 382).[22] Certainly, the discourses of disability studies and enhancement technology both reveal significant overlap with this early nineteenth-century amalgam of scientific speculation and radical political philosophy. There is an emancipatory social thrust to debates over whether people with autism represent an evolutionary step toward a new form of cognition, as to contemporary discussions of the relation between prosthetic and enhancement technologies, and even the possibilities of contemporary stem cell science (Cooper 2004).[23]

Indeed, this new medical and social openness to biological plasticity may perhaps explain the return to medicine by scholars in disability studies. "It seems to me dangerous for disability studies to neglect medicine and new biomedical technologies," Tom Shakespeare observed in 2005. "Disability studies has valuable work to do: distinguishing between appropriate and normalizing medical therapies; challenging doctors' tendency to define a disabled person's health and status totally with reference to their impairment; analyzing the potential, and dangers, of new treatments—such as stem cell therapy, gene therapy, [and] pharmacogenetics" (Shakespeare 2005, 145).[24] Shakespeare's observation suggests that a mindful investigation of teratology transformed may reveal new possible responses to the persistent parallel problems of disability studies: inadequate attention to the lived experience of disability within contemporary medicine and a correspondingly inadequate attention to medicine within some contemporary disability studies scholarship.

The [Pluri]potent Egg

In "The Triumph of the Egg," the teratology-obsessed father ends up with egg on his hands as he tries to demonstrate that "no one knows more of eggs than I do" (Anderson 1920, 13). Anderson makes the interesting decision to literalize the father's shame at failing to put his embryological knowledge to use, perhaps reflecting the stigma that attaches to the abnormal in his era. In contrast, the protagonist of the Australian film Yolk (2007) also ends up with egg on her hands as she searches for knowledge, yet for her this is neither shaming nor defeating.[25] This short film by Australian filmmaker Stephen Lance premiered in the United States at the 2008 Sprout Film Festival, a New York City event designed to raise the profile of people with developmental disabilities as subjects and performers by screening films of all genres featuring this population. The film concerns Lena, a fifteen-year-old girl with Down syndrome, who is curious about her newly awakened sexuality. Lena has been given a hen's egg to care for—a common exercise in biology classes in the Australian school system—and as Lena cares for the egg, we watch her develop sexually and socially.[26] Like *Days of Antonio*, *Yolk*'s aesthetic vision incorporates an activist agenda. Part of a broader effort to improve the lives of people with disabilities, it was filmed with the cooperation of the Down Syndrome Associations in New South Wales and Queensland, Australia. Like Browning's *Freaks*, it practices inclusive casting, featuring Audrey O'Connor, a young woman with Down syndrome, in her first professional acting role. And like Anderson's "The Triumph of the Egg," *Yolk* follows an adolescent through her sexual and social coming to maturity.

Because of her cognitive disability and because those who care for her assume that she is unable to come to grips cognitively with her own sexual maturity, Lena lacks the preparation to understand new feelings when she finds herself attracted to a neighbor boy. Her mother does not answer Lena's requests for information but instead merely parries Lena's question "How do babies come out?" with a question of her own: "Have you been hanging around with those boys? Have they said anything to you?"

Lena is a resourceful young woman, however, and quite capable of independent thought, so she begins a campaign to acquire the knowledge she desires. She steals *The Joy of Sex* from the local bookmobile and hides it under her bed, consulting it later that night alone in the bathroom. The next day, when she meets her crush at the bus stop, she leads him away into a woody glade. The conversation that ensues subverts the forces of stereotyping and stigma in film while representing Lena's ability to do so in life:

> "What's wrong with you?" the boy asks.
> "Nothing. What's wrong with you?"
> "Can I ask you a question? Why do you always carry that egg around?"
> "It's a baby."
> "Is it yours?"
> "Yes."
> "That's really sweet. Can I see it?"

The boy holds it, and as he does so, Lena reaches out and strokes his hand. But with that gesture, she has gone too far. The boy retreats and she is alone.

After walking home through the darkening forest, Lena finds her mother angrily chopping celery in the kitchen, back to the camera, with *The Joy of Sex* beside her on the counter. (That the viewer shares Lena's perspective in this shot reinforces the film's disability-friendly perspective.) "You come straight home!" Lena's mother yells. "I was worried sick! This wasn't on your [library] card. You stole it? I'm going to make you take it back." The camera focuses on Lena's hands as the mother scolds. She is still holding the egg, and her fingers

FIG. 14. "Lena and the Egg."
Used by permission of Stephen Lance

tighten until it is crushed, and yolk oozes between her figures. Lena retreats to the bedroom and sits on the bed, looking down at the mess of egg yolk and shell in her hands.

"It's dead," she tells her mother.

"It's not dead," is the mother's confusing reply. "It wasn't really a baby any way. We can just get another one out of the fridge and nobody will know the difference."

Although left with egg on her hands, Lena avoids the shameful defeat felt by the father in "The Triumph of the Egg." Instead, in its closing sequence, the film offers an alternative representation of disability that neither views Lena as a freak nor experiences anomaly and difference as monstrosity. When the scene opens, a somber Lena rides with her mother and siblings in the family van on the way to the library bookmobile, a replacement egg (from the fridge) rolling around precariously on the dashboard. The van pulls up at the bookmobile and Lena gets out:

"What are you going to say to them?" her mother asks.

"Say?" Lena replies, coolly, keeping her own council.

"I'll wait here for you," says her mother.

When Lena returns to the car we see she has not followed her mother's orders, for she hugs the purloined copy of *The Joy of Sex* close to her, snuggled under her coat. As the van pulls away from the library, we get a shot of Lena gazing out of the window with a dreamy smile. The scene then shifts to a closing image: Lena sprawled on deep green grass spangled with red and white flowers, gazing straight at the camera with her head resting pillowed in the lap of the neighbor boy.

In the film's ambiguous closing image, it seems at first that Lena has had her way. Despite the restrictions imposed on her by family and society, she has managed to acquire sexual experience—and with the boy who has been her longtime crush. But the soft focus of the scene, with the deep green grass and proliferation of flowers that pillow the lovers, soon cues us in to the fantasy nature of the image. We realize that we are squarely in Lena's mind as she formulates what sexuality means to her and about her. The message this concluding image conveys—to Lena and to the viewers—is an emancipatory one. Sexuality is more than just reproductive error, a hen's egg developing wrongly into a baby. It is a human relationship experienced and explored in terms meaningful to her. We could argue that the bit of genetic information in Lena's genome that is trisomy 21 is a difference that makes no difference; her Down syndrome is simply a variation, not pathology.[27] Yet that is too simple. Down syndrome has not just social consequences but medical ones, which can extend from congenital heart disease to developmental delay and mental retardation.

The vision of *Yolk* is more complex than either the moral, medical, or social models alone for understanding disability. We might instead find in the title of *Yolk* the film's central metaphor: the nourishing internal life force of Lena's imagination, which develops in her the self-reliance and determination necessary to navigate the challenges—social, conceptual, and physiological—of her impending sexual awakening.[28] So artists have drawn on the image of the chicken and the egg—in films such as *Yolk* and *Freaks*, short stories such as "the Triumph of the Egg," and plays such as *Days of Antonio*—to represent and explore the complexity of the disability experience.

Epidemic

"Fox-lox said, 'Come along with me and I will show you the way.'"

–Anon., "The Story of Chicken-licken"

emember the children's story in which Chicken-licken is hit on the head by a falling acorn? She turns it into a global catastrophe, racing around to warn all her friends, "The sky is falling!" (Anon. 1914). This story came to mind one morning in 2005 as I was writing about issues of risk and safety. That night I dreamed about a live-bird market: *There were exquisite small parrots and large macaws and pigeons. I was with a class and I told everyone it was urgently important that they wash their hands and clean their clothes before coming home with me because of the danger of passing on the avian flu to my chickens.*

Avian flu as a threat passing from humans to chickens? I awoke amused at how I translated Chicken-licken's panic into concern for my own chickens, the dream-work blithely rebutting the anxiety sweeping the nation over the possibility of a human epidemic of H5N1, the highly pathogenic avian flu. But I also woke convinced that this little children's narrative crystallizes elements of human awareness—folk wisdom of the Richard Hoggart or Raymond Williams variety—that we have lost as our agriculture and medical science have become targets of scientific management strategies designed to maximize safety by monitoring and minimizing risk.[1] As I worried about risk, safety, and the avian flu, the story of Chicken-Licken made me pause a moment to consider what constitutes risk, where risk comes from, and who is authorized to define, manage, and respond to it, as well as what constitutes safety, where we can find it, and what bargains we make in search of it.

Long years since I first heard it as a child, elements in the Chicken-licken folktale still seemed remarkably familiar to me in 2005, during the heyday of the avian flu panic. Among the many barnyard species that figure in the story—chickens and turkeys, geese and ducks—it is Fox-lox who causes the *real* trouble. For when the birds encounter the sly fox in the wood, as they are going to the king to beseech his aid in the "catastrophe" they imagine with that falling acorn, the fox offers to act as their guide: "Fox-lox said, 'Come along with me and I will show you the way.' But Fox-lox took them into the fox's hole, and he and his young ones soon ate up poor Chicken-licken, Henny-penny, Cocky-locky, Ducky-lucky, Draky-laky, Goosey-loosey, Gander-lander, and Turkey-lurkey, and they never saw the king to tell him that the sky had fallen!" (Anon. 1914, 15)

During the winter and spring of 2005–2006, the television news playing every day at the gym replayed this old story in a new form. Announcers warned of a newly intensifying form of avian flu, known as H5N1, or high pathogen avian flu, whose vectors were waterfowl and—it was feared—farm fowl. This new flu threatened to jump from birds to human beings, creating an epidemic that would necessitate fast and widespread government intervention. Amid the alarm and fear, Fox-lox (or rather Fox TV and their ilk) portrayed itself as guide, disseminating, defining, and even purporting to manage the nature and meaning of the crisis.

The Centers for Disease Control and Prevention (CDC) seemed to play the role of Chicken-licken when it offered this analysis of the situation: "Avian influenza is very contagious among birds and can make some domesticated birds, including chickens, ducks, and turkeys, very sick and kill them. . . . If H5N1 virus were to gain the capacity to spread easily from person to person, an influenza pandemic . . . could begin. . . . Experts from around the world . . . are preparing for the possibility that the virus may begin to spread more easily and widely from person to person."[2] The CDC message was clear: just like Chicken-licken and her friends, American citizens faced with the risk of avian flu must look to government and the media to define the risks the nation faces and the path to safety. Those experts tell them that danger lies in wild birds and back-yard chickens rather than large-scale poultry corporations.

Another form of awareness is generated by the expert knowledge circulat-ing around H5N1, where discourses of race, economics, and nationality con-verge to frame the meaning of risk and safety. To discover this infectious new awareness we must go not to the experts but to popular literature, whether the "airport paperback categories of the gothic and the romance, the popular biog-raphy, the murder mystery . . . the science fiction or fantasy novel," or the satiric newspaper *The Onion*, where we will find expressions of this new mode of think-ing (Jameson 1983, 112): "So, basically, the CDC doesn't have the first inkling of what to do about a potentially explosive form of flu that infects ducks and chickens," said Fox News science, health, and epidemics commentator Mary-linne Kent. "Given the popularity of these two birds as a food source among Asians, and the fact that we have no idea how many undocumented Asians have settled illegally in our nation, the potential for danger is extremely high."[3] Con-tagion has once again become a metaphor (see Wald 2008). In the *Onion*'s spot-on satire, Fox's racialized presentation not only hypes risk—it actually constructs it. The alien Other is baldly made into a criminal (an illegal alien) and then rendered even more dangerous by being assumed to be the illegal source of the epidemic. The health of the nation is endangered by criminality of those who cannot be accounted for or traced.

As "Chicken-Licken" with its acorn and the *Onion* satire with its "undocumented Asians" both reveal, expert knowledge about the production of life frequently functions not to inform but to obscure. While this has been given ample demonstration in relation to contemporary biomedicine, it is also the case, I argue, with the converging expert discourses that construct what risk and safety mean in terms of H5N1. As Ulrich Beck specifies, "*Risk* may be defined as a *systematic way of dealing with hazards and insecurities induced and introduced by modernization itself*" (a process that of course also involves the "hazards and insecurities" of—illegal—immigration) (Beck 1992, 21; emphasis in original). Risk, in Beck's analysis, is reflexive, a source of ever more economic productivity as it mandates increasing expert knowledge, rationalization, and surveillance. Crucially, risk is also a source of what Beck calls unawareness (including unawareness of the life circumstances of other alien bodies), a systematic production of ignorance that operates by defining knowledge more and more narrowly as the product of scientifically mediated institutions, so that laypeople are no longer confident about the wide range of common sense, tacit, everyday knowledge available to us (Beck 2000). The impact of risk may be imagined in global terms (aligning with the expert analyses of global corporations), but in reality risk has an unevenly distributed global impact. It exacerbates old inequities while producing new ones, such as the devastating financial losses faced by small chicken farmers when an outbreak of avian influenza in high-density poultry farms leads to a forced cull of outdoor birds or the increased rate of assembly-line injuries suffered by poultry workers when a global dip in chicken prices leads to a speed-up on the production floor. We are promised (false) security—healthy food—in exchange for surrendering our abilities to seek out the wider range of alternative analyses and responses such as those forms of knowledge either unknown to or not sanctioned by the CDC.

The very habit of seeing risk everywhere and enforcing specific measures designed to produce safety in response supports hegemonic scientific rationality. Safety discourse also defines what we come to think of as a state of security and what are the authoritative measures of the economic and social costs of achieving security; it authorizes expert knowledge (and in turn creates the benchmarks for such authorization); and, most important, it creates a particular form of unawareness on which this whole discursive apparatus of safety relies. This "double construction of unawareness" includes both the obstruction and rejection of other forms of knowledge and the "*denial of our inability to know*" (Beck, 2000; emphasis mine). To invoke Donald Rumsfeld: *We don't know what we don't know.*[4] Because we focus on scientifically defined risks and technologically mediated solutions, we are unable to grasp aspects of experience not subject to quantification. Because we are focused on increasing the biosecurity of existing poultry production units, we do not even consider the broader social, biomedical, and cultural consequences of raising genetically similar chickens in the stressful conditions of overcrowded, confined poultry houses. "The Story of Chicken-licken" and the *Onion* news report ("Nation's Leading Alarmists Excited about Bird Flu") both reveal the systematic process of producing unawareness that plays an essential part of the reframing of risk and safety accomplished by industrial poultry production.

Part of the process of rearticulating risk is repressing from public memory the fact that low pathogen avian influenza has been around for centuries, finding its historical reservoir in the bodies of wild waterfowl and making occasional forays into flocks of chicken or turkeys, such as the eruption of H5N2 virus in Pennsylvania poultry farms in 1983 (Greger 2006, 34; see also Davis 2005). There is significant consensus that the 1918 human influenza epidemic had a porcine source. Even as early as 1919, an investigator for the U.S. Bureau of Animal Industry named J. S. Koen published an article in the *American Journal of Veterinary Medicine* arguing, "[the] similarity of the epidemic among people and the epidemic among pigs . . . [suggested] a close relation between the two conditions for the 1918 [epidemic]" (Koen 1919, 468). Since such a claim had its own economic risks to the growing pork industry, there were strong disincentives to explore the possibility of a transspecies transmission of avian flu. There is, then, nothing new about avian flu. The pathogen has been with us for years—only the context has changed.

However, the entire history of the disease was reframed when a highly contagious and very lethal strain of this virus emerged in Hong Kong in 1997, gaining widespread media attention. This new variety not only killed chickens, but it also made the trans-species jump to human beings, killing three-year-old Lam Hoi-Kaw in May 1997 (Greger 2006, 32). The Hong Kong outbreak was eradicated by a systematic slaughter of more than 1 million chickens, applauded in 1998 by a joint proclamation signed not only by scientist experts on the influenza virus but also by the World Health Organization (WHO), acknowledging, "We may owe our very lives to their actions" (quoted in Greger 2006, 37). In 2004 another virulent outbreak of H5N1 avian flu spread across Southeast Asia, and the alarm generated by this outbreak triggered a media panic.

Stefan Lovgren's story in the online *National Geographic News* exemplifies this journalistic hysteria. Lovgren breathlessly reports that the virus was spreading not just from person to person but from species to species: "This year there have been 44 confirmed human cases of H5N1 flu in Thailand and Vietnam. Of these, 32 people died. There is not yet a vaccine for the disease. . . . Meanwhile the virus has undergone huge genetic changes and become even more pathogenic. It now affects not only birds, but also cats, pigs, and even tigers" (Lovgren

2004). The *National Geographic News* article is a masterpiece of doublespeak. The author gives ample space to the alarming announcement of Shigeru Omi of WHO that avian flu death estimates "of 2–7 million deaths were 'conservative' and that the maximum range could go as high as 50 million deaths." Then he reports wryly, almost regretfully, that despite these grim predictions, "the one thing that did not break out was mass panic" (Lovgren 2004).

The journalist's nose for news seems to be sniffing hard, indeed, wishing for a good outbreak of public hysteria to document for his readers. He cites a comment from UCLA virologist Michael Lai, "This alarmist warning is irresponsible in using this language to rouse the public's fear," but gives Dick Thompson, a WHO official in Geneva, Switzerland, the last word: " 'Are we scaring people? I don't know,' he said. 'But rather than springing on people some terrible event, it's better that they get emotionally ready for what they could face. We think a pandemic is coming. Nobody knows when. But it is good to get people prepared before it arrives.' " True to its orientalist heritage even in its title, Lovgren's piece, "Is Asian Bird Flu the Next Pandemic?" adopts the persistent racializing misnomer—the "Asian bird flu"—thus contributing to the racially inflected and pathologizing anti-immigration discourse swirling around the disease (Lovgren 2004).

That highly pathogenic avian influenza could jump from an avian species to humans, so that the disease could then be spread directly person to person the way conventional influenza does, was generally dismissed until 2005. Then a team led by pathologist Jeffrey Taubenberger and including scientists from the CDC and New York's Mount Sinai Hospital sequenced the influenza virus contained in tissue samples of victims of the 1918 influenza that had been preserved in Alaskan permafrost (see Greger 2006; and Davis 2005). In a *Nature* essay from October 2005, the team announced: "Here we present sequence and phylogenetic analyses of the complete genome of the 1918 influenza virus, and propose that the 1918 virus was not a reassortant virus (like those of the 1957 and 1968 pandemics), but more likely an entirely avian-like virus that adapted to humans. These data support prior phylogenetic studies suggesting that the 1918 virus was derived from an avian source" (Taubenberger et al. 2005, 889. Folk wisdom had registered awareness that would remain inaccessible to science for years: "Back in 1918, schoolchildren jumped rope to a morbid little rhyme: 'I had a little bird, / Its name was Enza. / I opened the window, / And in-flu-enza' " (Gregor 2006, 13).

Taubenberger's study, and one that followed it in *Science* in 2005 by Terrence Tumpey and others at the CDC, raised public fears not only because of the fatal conclusion but also because of the methodology the researchers had used (Tumpey 2005). Both research groups had recreated a strain of the virus from RNA fragments preserved by the permafrost and published the full genome sequence of the virus on the GenBank database. This material was now open-access, according to Jonathan Tucker of the Center for Nonproliferation Studies: "If someone wants to reconstruct the virus, says Taubenberger, 'the technology is available'" (quoted in Bubnoff 2005, 795).

In 2004 former U.S. secretary of health and human services Tommy Thompson described the avian flu as more of a threat than bioterrorism, but within a year the two threats merged as the oversight responsibility of the National Science Advisory Board led them to define avian flu as a bioterrorist threat to the nation. The U.S. National Science Advisory Board for Biosecurity convened a special meeting upon publication of Taubenberger's article. While they concluded that the benefits of such research "clearly outweigh the risks," they requested that Taubenberger and his fellow researchers add a passage to the manuscript stating that the work is important for public health and that it had been conducted safely.[5] However, as von Bubnoff observed in *Nature*: "Taubenberger admits that there can be no absolute guarantee of safety. 'We are aware that all technological advances could be misused,' he says. 'But what we are trying to understand is what happened in nature and how to prevent another pandemic. In this case, nature is the bioterrorist'" (Bubnoff 2005, 795). Taubenberger's recourse to the ready-to-hand rhetoric of terrorism, so constitutive of all U.S. discourse since 9/11, in response to criticism that his research methods are unsafe reveals the constructed nature of the concept of "safety," shaped as it is by multiple extrascientific factors.

Chief among these factors, of course, is the economic bottom line. In the decade of pandemic hysteria since 1997, the avian flu publishing industry boomed. Local and national government agencies including the Food and Drug Administration, the U.S. Department of Agriculture, and the Centers for Disease Control and Prevention issued warnings and informational pronouncements, and on May 9, 2006, ABC aired the made-for-TV movie *Fatal Contact: Bird Flu in America* (Pearce 2006). Indeed, science and entertainment combined forces when the movie became the subject of analysis by the Department of Health and Human Services on a government Web site with the same title as the film, "Bird Flu in America."[6] That avian flu was a good profit opportunity for businesses, the media, and even government institutions was amply demonstrated by one recent example: the program for the Fifth International Bird Flu Summit held in September 2007 in Las Vegas. This New-Fields conference brought "distinguished scientists, international health organizations and world leaders" together with "heads of the world's top companies to discuss how the world can survive an imminent pandemic."[7] In addition to keynote speeches by experts, there were breakout sessions exploring "Business Continuity Planning," "Emergency Management Services," "First Responders Law Enforcement/Police Department/Public Works," as well as the customary exhibition hall with its range of pharmaceutical, technological, and genomic offerings. All this was available to conference participants for the "super early bird" rate of US$1,850. Well worth it, according to the New-Fields conference Web page, where glowing testimonials from representatives of the United Nations, the U.S. Army, the U.S. Department of Defense, and the U.S. Department of Agriculture (USDA), as well as academics, were posted alongside those from representatives from Dow Biocides, F. Hoffman-La Roche AG, the U.S. Department of Homeland Security, Karl Hans-Fuchs Collective Protection Engineering, and the European Influenza Surveillance Scheme.[8]

The "Chicken" in "Chicken-licken"

Any discussion of the "monster at our door"—as Mike Davis has dubbed it with noir irony—must begin with something more prosaic and less media friendly: the transformation in chicken farming in the United States since 1900 (Davis 2005). Before the consolidation of industrial poultry farming in 1932 with the formation of the Institute of American Poultry Industries, most chickens were raised in the farmyard or the garden. In the preindustrial era, the meaning of risk in any agricultural context was straightforward: the livestock could die, leaving the farmer with no animals to sell for meat and no profit. Even in chicken farming, the risk that chickens might stop laying or might die threatened the extra income generated by the backyard flock. Safety, too, was easy to define: it referred to the chicken's state of freedom from predators and access to sufficient food and water, and to the farmer's possession of adequate production and distribution avenues to generate adequate income from the chickens. Such safety depended on the individual chicken farmer's skill in managing the health and productivity of her layers and meat birds, as well as business sense: good poultry husbandry, in short. The curiously outdated term is significant: as we saw earlier, farmhouse-based chicken raising characteristic of the United States in the early to mid-twentieth century was a deeply gendered activity. Most chicken farmers were women, and the backyard flock was typically the responsibility of the farm wife or daughters, who managed the flock's illnesses with medical practices traded from farm to farm. Chickens insulated the farmer against economic losses caused by crop failures or other livestock losses, providing eggs and meat for the family table as well as supplying the farm women with precious "egg money" held separate from the main sources of farm income (Sachs 1996). For farmers, sharing veterinary advice with each other was a very simple strategy of communal risk management and economic survival.

Just about the time the term "animal husbandry" entered the lexicon, around 1915 to 1920, this reliance on nonexpert or folk wisdom in chicken raising and doctoring began to change, reflecting (in part) the influence of an English-born veterinarian named Joseph Edward Salsbury (Sawyer 1971, 26, 36, 112).[9] "Doc" Salsbury professionalized chicken medicine, radically transforming chicken farming in the process. He provided veterinary advice and mail-order patent poultry medicines. Moreover, he sold his services as a "specialist on poultry diseases" (63). He began offering annual poultry short courses to the general farming public. The first one, held on October 12–21, 1931, combined morning talks on specific diseases by well-known speakers with afternoon sessions covering laboratory work and clinical medicine. The no-fee courses attracted leaders in poultry farming, poultry specialists, and poultry supply dealers. By midcentury, more than ten thousand people had attended the "Dr. Salsbury poultry disease schools." "These men, to a very large degree, represented the basic school of knowledge which would help lead the commercial poultry industry into being" (64). Certainly there might have been a woman or two who attended the short courses, but given the fact that they began as invitation-only events and expanded from that to the foundation of the poultry industry, the attendance at these short courses was overwhelmingly male. Salsbury's system for research, marketing, and educational consultation not only paved the way for the academic field of poultry science but it also anticipated the growth of poultry raising as a profit center linking the vertically integrated poultry industry (that is, one that goes from incubation to processing) to pharmaceutical companies specializing in poultry medicine.

As Salsbury's veterinary business was growing in the early twentieth century, land grant colleges in the United States were making poultry education part of their mission. The University of Connecticut led the way in the establishment of poultry science departments (1902), followed by such recognized chicken farming states as New York (1907), Washington (1918), Massachusetts (1920), Indiana and Michigan (1921), and Pennsylvania and North Carolina (1924) (Thaxton et al. 2003, 305). Educators and extension educators began systematizing the practices of men involved in breeding, farming, and exhibiting, such as Joseph McKeen of Omro, Wisconsin, who developed the Buff Wyandotte variety; Joseph Wilson, who pioneered shipping day-old baby chicks; and Isaac K. Felch, "credited as being the first poultry judge to make his entire living from the poultry industry" (Skinner 1974, 38).

Animal husbandry was starting to become a masculine field. No longer were women the primary chicken farmers, keeping their chickens healthy with their own home remedies. Chicken raising had been gendered female, but that era was coming to an end, as anthropologist Deborah Fink learned in her ethnographic study of an Iowa region she called "Open Country": "So completely were chickens associated with women that older Open Country people frequently categorized chicken chores as housework" (Fink 1987, 49; see also Sachs 1996). The extension educators of the interwar era who hoped to turn poultry raising into a scientific practice suitable for the modern male farmer faced a challenge. Before 1940, activities associated with poultry raising and egg farming were still stigmatized for men precisely because of their historical association with women.[10] "Chickens were the classic bane of men, but this did not keep them from being the most common enterprise on pre–World War II Iowa farms" (Fink 1987, 49). Yet at the end of World War II, chicken farming began to appeal to men, as E. B. White explained with mock earnestness: "Right now the hen is in favor. The war has deified her and she is the darling of the home front, feted at conference tables, praised in every smoking car, her girlish ways and curious habits the topic of many an excited husbandryman to whom yesterday she was a stranger without honor or allure" (White 1944, v). The sudden increase in male chicken farmers reflected a redirection within the field of agricultural extension education: "American women's exclusion from egg and poultry production resulted from a conscious policy decision on the part of agricultural program planners. . . . Rather than encouraging and sheltering women's poultry production, the postwar extension service ceased to include women . . . unless their husbands were also participants" (Fink 1987, 135).

The new academic field of poultry science linked research in basic sciences with applied research on poultry production and management, producing (predominantly male) avian scientists and managers for the growing poultry industry. Although women were not entirely excluded from the poultry industry, their roles changed dramatically. As the ownership of the industry was consolidated in larger and larger corporations, class differentiation joined gender sorting. Poultry workers for this new industry were not only frequently female but were also recruited from poor, disenfranchised, migrant and immigrant populations.[11]

As the industry increasingly turned to technological innovations to manage the risks of poultry production, it also began to incorporate technologically mediated definitions of risk and safety. This was certainly the case with incubation. Although incubation was one of the oldest technical interventions in chicken raising—a practice dating back to the ancient Egyptian practice of incubating chickens in large clay rooms heated with manure-fired ovens—the introduction of a modern electrically heated incubator in 1923, developed by Ira M. Petersime, launched a debate about the ethics and effectiveness of such practices (Sawyer 1971, 26). Traditional poultry breeders argued that it was "morally wrong to hatch chicks artificially." The American Poultry Association, formed in 1873, itself joined the debate, waging a campaign that "set out methodically to tell the American farmer the quality of artificially incubated chicks was inferior, and that it would be impossible to transport them successfully" (27). However, breeders who were selling these new "artificial chicks" countered the association by marketing their product as more scientifically advanced and thus better. By 1916, when the American Poultry Association gave way to the new International Baby Chick Association, scientific rationality and industrial efficiency had prevailed. Chicks could now be mass incubated and delivered to customers via the U.S. mail. One of the many contemporary handbooks on poultry raising, *Making Money from Hens* (1919), characterized the result as a safer and more effective means of distinctly American chicken production:

> The development of manufacturing enterprise, coupled with our Yankee inventive genius, conceived and rapidly developed the efficient artificial incubators we know to-day, until at the present time they are far superior to the hen, in that they hatch better. . . . With incubators the time of hatching is not subject to the whims of the hen, but is absolutely under the control of the efficient poultryman.
>
>   Chicks artificially hatched and reared are not subject to the parasites and disease contamination they are bound to contract to a greater or less extent when running with hens.   (Lewis 1919, 59–60)

The discourse of risk folds almost seamlessly into the discourse of safety, inflected with gender and nationalism even at this early stage in the industry's development. The risk of a poor hatch ratio is countered by the safety provided by the efficient Yankee incubator, just as the risk of exposure to parasites and disease is mitigated by keeping the artificially hatched chick indoors. Thus, closed in, under the control of the efficient "poultryman," the chick is safe from even the small degree of danger it would encounter "running with hens" (60). The rhetorical redefinition arrests in medias res the transition from backyard to industrial poultry farming: the hen-and-nest method of chicken incubation and growth documented in Elmer Boyd Smith's 1910 "Chicken World" is redefined as risky (unsanitary and uncontrolled) while the artificial incubator and confined growth are redefined as safe. Moreover, safety becomes technologized, medicalized, racialized, and nationalized: it is conceived of as freedom from disease and parasites associated with foreign exposure, available to chicken raisers thanks to "Yankee inventive genius" (59).

By the time of World War II, these new technologies and modes of expert knowledge had solidified into what Deborah Fitzgerald has described as the ideal of industrial agriculture, characterized by "timeliness of operations, large-scale production sites, mechanization, standardization of product, specialization, speed of throughput, routinization of the workforce, and a belief that success was based first and foremost upon a notion of 'efficiency'" (Fitzgerald 2003, 5). Risk had been reframed, from something controlled by the individual farm wife (who raised backyard chickens to manage the economic risk of farming) to something controlled by corporate risk management departments through structural innovations in poultry farming. Safety, too, had been redefined as the product of scientific rationalization and expert control, with distinctly economic overtones. Cost–benefit ratios ruled. By the 1960s, the poultry industry had adopted the conglomerate model, in which one corporation has subsidiary companies that perform the various stages of poultry production in different places: hatchery, growing houses, processing houses, feed suppliers, and distributors. The companies provided the inputs (eggs, chickens, feed, and medication) while the contractors supplied the housing, labor, energy, and especially the risk (whether of flock failure or price decline) (Bugos 1992, 146).

With this new vertical integration, risk and safety were segmented. Risks were borne primarily by the poultry workers, both the contractors who had to provide cash and infrastructure for the poultry raising and the line workers in poultry plants, whose bodies were constantly at risk of injuries ranging from blindness to amputations. Safety became the achievement of experts in the poultry corporations who developed strategies for preventing flock failure or theft of intellectual property. These included the discovery of vitamin D, the use of electric feeders, and the application of commercial vaccines against poultry illnesses, all of which made it possible for birds to be raised inside on an unprecedented scale, thus (it was thought) keeping them safe from pathogens. As knowledge about chicken breeding was gradually redefined from public to corporate property, a corporation's genetic property could also be made safe from theft, dilution, or imitation of the breed. The poultry industry increased its economic security by hybridizing to increase egg yield or meat production, developing birds with specific qualities for designated markets (see Thaxton 2003; and Boyd 2001). So a new three-step breeding practice produced a uniform, standardized chicken that could not be replicated by farmers but had to be purchased yearly from the breeder. As Glenn Bugos notes, "Bred directly into the hybrid chick was the means to keep them from being illegally reproduced" (Bugos 1992, 144). If risk had been curtailed, so had profit—it was principally reserved for those who owned the technology rather than those who raised, tended, and slaughtered the poultry.

The risk-management strategy of increasing control over poultry as a form of intellectual property led to innovations in both the hatchery and the scientific laboratory. A premier instance of such innovations was at the Cobb Corporation, founded in 1916 as "Cobb's Pedigreed Chicks" by Robert C. Cobb Sr., a Harvard graduate.[12] By 1926 Cobb had become known as New England's major breeder of Barred Plymouth Rocks. Aware that consumers preferred all-white birds since no dark quills were left in the skin when the poultry was processed, Cobb responded in 1947 by breeding its all-white hens with the white male birds produced by another corporation, Vantress, to produce a fully white meat bird. In 1974, just as Cobb was purchased by the Upjohn Corporation, its partner breeder, Vantress, was simultaneously purchased by Tyson Foods. Twelve years later, Tyson, the owner of the Vantress pedigree line, and the Upjohn Company, owner of the Cobb 500 breeding program, the systematic practice of controlled breeding to produce chickens with certain desired characteristics, merged to become the Cobb-Vantress Corporation, with corporate headquarters in Siloam Springs, Arkansas, and subsidiaries in Africa, Southeast Asia, and South America. Apt indeed that until the turn of the twenty-first century the corporate logo of Cobb-Vantress should show a rooster whose eye is the globe.[13] No chicken, or chicken-related technology, it would seem, was beyond the Benthamite eye of Cobb-Vantress. Yet like all successful corporations, Cobb-Vantress is above all flexible; by 2010 the Cobb-Vantress Web site had responded to the new interest in sustainable farming with the introduction of their new product, the "Cobb-Sasso 150," which the corporation describes as: "the natural choice for consumers interested in slower growing, colored chicken. The broiler's robust health and well-being are ideally suited to traditional, free range and organic farming as well as less intensive indoor production. The mating of a rustic brown female and white male give the broiler a distinctive look and excellent growth performance."[14]

Once the price of chickens dropped due to a glut in the chicken market in the mid-1960s, there was little profit in selling whole birds, and chickens were increasingly designed for the competitive global market, in which the most money was to be made by adding value after processing. This technique was inaugurated by Cornell University poultry science professor Robert Baker, who laid the groundwork for chicken nuggets by inventing a deboning machine that took all of the meat off the chicken carcass and by developing a variety of ways of reshaping the extracted meat: into dinosaur-shaped nuggets, chicken bologna, chicken pastrami, and chicken ham (Martin 2006). The resulting interest in adding value created incentives for producers. Cobb-Vantress's most recent corporate product, the Cobb 700, was specifically designed for the South American and European markets, where "processors supplying the high meat yield, deboning and added value markets" were soon showing great interest in this new product. It was framed as a major improvement even within the Cobb-Vantress line: "Most important of all, breast meat yield of the Cobb 700 broiler had improved by a full one percent over the Cobb 500 broiler!"[15] This new bird would keep corporate profits safe, supplying postprocessors with the meat they needed to make the largest profit possible, whether they were producing precut frozen breast portions or deboned, ground, reshaped, and processed chicken products. Such "industrial" birds, designed as sequentially numbered "models," are bred, brooded, and battery-raised (that is, indoor rather than free range) in high density throughout the world.[16] In a global food industry pressured to compete with increasingly higher yields and lower costs, the U.S. (factory) model of poultry farming drives out many of the small local poultry raisers who are less insulated from the risk of price shifts and distribution difficulties (Sachs 1996).

The dominance of risk management strategies reaches beyond poultry corporations into the realm of avian science as corporate laboratories work with the pharmaceutical industry in the production of engineered birds. Even seemingly pure scientific research has corporate tie-ins, though they may be obscured by the discourse of unfettered scientific progress. In 2004 the International Chicken Genome Sequencing Consortium completed sequencing the genome of *Gallus gallus*, the red jungle fowl from which all domestic chickens have descended (ICGSC 2004). This accomplishment, two genomics researchers explained in an article in *Nature*, would benefit "agricultural researchers attempting to breed the most productive strain by recognizing links between DNA sequences and attributes such as egg production" (Schmutz and Grimwood 2004, 679). What the authors failed to mention is the fact that not *all* information obtained through processes developed with monetary support from the poultry industry is available for public analysis or use. Instead, it is considered to be the industry's property.[17] Although the Dutch poultry company Hybro calls itself "your partner in breeding," the information produced by company-funded technical processes is *proprietary*—it is not available to the general researcher or chicken farmer without permission. Scientific progress is held hostage to corporate profits. No profits without science; no science that is not exclusive corporate property.

As one commentator observes, corporate poultry science could even use transgenics to make a profit off the avian flu, melding the goals of economic and biomedical safety: "Agribusiness companies stand to reap huge gains in the event that scientists at Cambridge University and elsewhere are able to replace the entire world chicken population with genetically-engineered chicks allegedly resistant to H5N1 virus" (Engdahl 2005).[18] In risk lies the opportunity for profit, whether by breeding "better" chickens or by cashing in on the bioengineering possibilities generated by avian illness.

Defining and Managing Risk

Issues of economic and epidemiological risk converge at the point of avian flu to produce a new rhetoric of safety. The major players in the global poultry industry have responded to concerns about avian flu by assuring customers that their products are safe precisely because they are highly engineered and scientifically monitored. So Cobb-Vantress explains: "A cornerstone of the company's success is the adherence to strict internal bio-security and safety standards throughout all operations. Facilities are designed to accommodate the highest bio-security and safety standards to ensure consistent delivery of quality product to customers globally. Bio-security and safety know-how and experience gained internally is often passed onto customers as an added benefit of dealing with Cobb as a supplier of breeding stock."[19] Tyson, a major customer of Cobb-Vantress's chicken lines, takes a very narrow view of the sort of know-how to be passed on to its consumers. This is a far cry from the era when chicken raising depended on farm and folk wisdom, the era before Doc Salsbury began the systematization of poultry farming. Tyson's Web site promises "products, recipes, and peace of mind," thus effectively isolating concerns about safety to the gendered realm of cooking and consuming.

A low-pathogen strain of avian influenza evolves into a high-pathogen strain precisely because it has passed repeatedly through the great numbers of chickens held in high-density confinement poultry houses. The sheer number of potential mutations available to a virus in such circumstances elevates the likelihood that a new strain will appear, one no longer limited to the bird-to-human infection process. This is the fear, then: that such a high-pathogen strain of avian influenza *with the new capacity of direct human-to-human transmission* will emerge from the crowded, filthy conditions of the factory farm. The culprit in that case will clearly be the poultry industry, whose practices may well have accelerated the movement of the virus through the avian population.

Veterinarians and conservationists identify two practices common to industrial poultry production that have contributed to the seeming speed with which H5N1 has traveled from its emergence in Southeast Asia to its current outbreaks in Europe and Africa: the global shipping of day-old chicks and the practice of intensive battery-raising of young adult birds. "The chick trade 'has made the chicken the most migratory bird in the world'" according to Adrian Long of BirdLife International (McNeil 2006). Mike Davis has laid the blame for the intensive battery-raising of young adult birds on the international poultry industry, which he argues has deliberately framed "biosecurity" by distinguishing between the so-called high-security, high-volume growers and the purportedly unsafe and unsanitary methods of backyard poultry growers (see Davis 2005; and Gregor 2006). Indeed, rather than adopting the misnomer "Asian bird flu," it might be more accurate to call this the "free trade flu," since we can attribute the increasing risk of an explosive outbreak of H5N1 to the global movement of day-old baby chicks from U.S. corporate hatcheries to contract growers overseas coupled with the exportation of U.S. methods of high-volume poultry raising. The rapid mobility of poultry has introduced a potentially new element into the international transportation of chickens: pathopolitics. "When a new virus gets into a barn packed with thousands of young chickens that have been genetically selected for their plump breasts rather than their ability to survive in the wild, it leaps from bird to bird, mutating slightly each time, and sometimes morphs into a lethal strain—just as the 1918 Spanish flu was believed to become more deadly as it passed through crowded American military camps during the cold winter of 1917" (McNeil 2006).

In fact, pathopolitics aside, the industrial poultry industry raises issues of risk and safety even before we consider the threat of avian flu. As it is currently structured, conventional large-scale poultry production poses a risk to the physical and emotional health of human beings as well as the chickens it produces. A January 2005 report by the U.S. Government Accountability Office

(GAO) documented that the young, male, and/or predominately Hispanic workers had rates of injury "among the highest of any industry." These injuries included not only "cuts, strains, cumulative trauma caused by repetitive cutting motions," but also "injuries sustained by falls, more serious injuries, such as fractures and amputation," and illnesses caused by "exposure to chemicals, blood, and fecal matter" (GAO 2005, 21–22). It is worth noting that even the GAO got precious little cooperation from the poultry industry in carrying out the investigation leading to this report.[20] The very characteristics of contemporary poultry farming—in which the chickens are subject to overcrowding, stress, filth, lack of sunlight, and induced immunosuppression from selective breeding and monoculture—pose a health risk for human beings as well. As Greger points out, "stressful, overcrowded confinement in industrial poultry facilities facilitates immune suppression in birds already bred with weakened immunity, offering viruses like bird flu ample opportunities for spread, amplification, and mutation" (Greger 2006, 214).

Because it ignores such local and specific problems for a global and ideologically more acceptable danger, the bioterrorism metaphor invoked by Taubenberger to describe the avian flu is, like all metaphors, enabling as well as merely descriptive. In 2003 the Pentagon proposed a program called "the Futures Markets Applied to Prediction (FutureMAP)." This initiative of the Pentagon "would have involved investors betting small amounts of money that a particular event—a terrorist attack or assassination—would happen." Bad press on Capitol Hill killed this initiative, according to a CNN report. "Tom Daschle (D-SD) noted, 'I can't believe that anybody would seriously propose that we trade in death,' while Senator Barbara Boxer (D-CA) observed, 'There's something very sick about it'" (Courson and Turham 2003).

No such objections seem to shadow the Iowa Health Prediction Market. Framing itself as "a step beyond disease surveillance," this initiative of the University of Iowa invites health care workers to wager donated money on the risk of an avian influenza outbreak, essentially offering them a risk-free investment in risk. Its Influenza Prediction Market uses the profit motive to encourage "physicians, nurses, pharmacists, clinical microbiologists and epidemiologists" to share the information that may enable the forecast of influenza activity before an outbreak actually occurs. Giving each health care worker/trader an "education grant of $100 with which to trade," the market gauges the "consensus belief" about the likelihood of an influenza outbreak. The Influenza Prediction Market Web site explains that it views the Influenza Prediction Market as a "supplement that can quickly aggregate expert opinions based on existing surveillance information. . . . The probabilities generated by this market could help policymakers and public health officials coordinate resources, facilitate vaccine production, increase stockpiles of antiviral medications, and plan for allocation of personnel and resources."[21]

Antiterrorist anxieties also seemed to be behind an initiative explicitly aimed at increasing food safety, the National Animal Identification System of the USDA. This plan, a USDA pamphlet explained, was created in response to the "increasing number of animal disease outbreaks . . . reported around the globe" as well as the "single cow that tested positive for bovine spongiform encephalopathy (BSE) in the United States in December 2003." The ornate arrangements proposed under this system included seven-character IDs for individual livestock producers, fifteen-character IDs for individual animals, and thirteen-character IDs for groups of animals. Although the program proposed to "enhance foreign animal disease surveillance, control, and eradication" and thus to "improve biosecurity protection of the national livestock population" by issuing "official identification for animals in interstate or international commerce," it met such a storm of criticism that it was withdrawn.[22]

On February 5, 2010, the USDA issued a news release: "After concluding our listening tour on the National Animal Identification System in 15 cities across the country, receiving thousands of comments from the public and input from States, Tribal Nations, industry groups, and representatives for small and organic farmers, it is apparent that a new strategy for animal disease traceability is needed," said Secretary of Agriculture Tom Vilsak. "I've decided to revise the prior policy and offer a new approach to animal disease traceability with changes that respond directly to the feedback we heard." This new approach would, according to Secretary Vilsak, apply only "to animals moved in interstate commerce"; would "encourage the use of lower-cost technology," and would be "administered by the States and Tribal Nations to provide more flexibility" while also being "administered transparently through federal regulations and the full rulemaking process."[23]

Despite the constructions of risk and safety implied by such cloak-and-dagger strategies, the crucial danger with avian flu is not the possibility that the virus will jump from birds to humans or cause a pandemic. Rather, to avian diagnostic pathologist Dr. Patricia Dunn, the real impact of the avian flu lies in its potential to compound already existing global economic and health disparities. Thus the greatest risks are to the emotional health of farmers and their families who may lose their flocks to an outbreak, thus losing livelihood, and to nutritional health if people lose their access to chicken meat and eggs, one of the few easily accessible sources of protein for people living on a subsistence level throughout the world.[24]

Coda: H1N1

By 2009 the species-jumping influenza to worry about was H1N1, not H5N1. The avian flu, focus of fear just three short years earlier, had fallen far short of the feared public health disaster. Although a number of flocks of so-called outdoor chickens were euthanized in Europe and Asia, no widespread transfer of the virus from wildfowl or domestic chickens had occurred. The cumulative total of cases of H5N1 in human beings that had been reported to the World Health Organization since 2003 was 471; the cumulative total of deaths was 282 (WHO 2010).Time had moved on, and what had first been called a "global swine flu pandemic" was now causing serious concern among public health officials, government agencies and NGOs, and, inevitably, the media (Schmidt 2009, A394). Cases of this virulent new strain of the flu, normally found in pigs, had been documented in human beings in more than 170 countries. The history and significance of the swine flu epidemic is far too complex to sum up here. Instead, I want to close by considering how two aspects of the story of this new outbreak of H1N1 complicate the story of Chicken-licken with which I began: the conflict it revealed between practices of industrial farm food production and techniques of mass vaccine production, and the connections between species it both invoked and repressed.

The avian and swine flu epidemics share a familiar pathopolitics: a viral mode of information transmission that spurs panicky rumors. In the case of swine flu, rumors in the blogosphere first linked the outbreak of H1N1 in humans to industrial agriculture. Early in 2009 the Huffington Post ran a piece by David Kirby claiming that the Mexican origin of the swine flu outbreak could be traced to the "hundreds of industrial-scale hog facilities that have sprung up around Mexico in recent years, and the thousands of people employed inside the crowded, pathogen-filled confinement buildings and processing plants" (Kirby 2009). Kirby painted a vivid picture of the susceptibility of even supposedly "sterile" confinement operations to contamination and transmission of influenza viruses, citing the Pew Commission Report on Industrial Farm Animal Production that appeared a year earlier (Pew Report 2008). The Pew report to which Kirby referred assessed the effects—on public health, the environment, the welfare of animals, and the state of rural communities—of the industrial-scale model of farming animals for meat. Although acknowledging that "antimicrobial resistance, zoonotic disease transfer to humans, and occupational and community health impacts that stem from the dusts and gases produced by IFAP [Industrial Farm Animal Production] facilities—are not unique to industrial farm animal production," the report was unambiguous in its verdict. "It is the size and concentration of IFAP facilities and their juxtaposition with human populations that make IFAP a particular concern" (19). It recommended moving "towards sustainable agriculture" (50).

When the H1N1 panic first hit the news, particularly singled out for suspicion by the blogosphere was the Granjas Carroll pig farm in La Gloria, Mexico. This farm, partly owned by the U.S. corporation Smithfield Foods, kept nearly a million swine and their odorous effluent in an area near residential neighborhoods.[25] Four days after the Huffington Post ran this story, Curtis Brainard pushed back at the rush to indict factory farming for the spread of this novel flu, acknowledging on the *Columbia Journalism Review* blog "The Observatory": "So far . . . there is no evidence of a direct connection between the farm and the swine flu virus. But there are reasons both to suspect and doubt that such a connection exists, and this has led to sporadic arguments among reporters covering the outbreak about the line between asking tough questions and jumping to conclusions" (Brainard 2009). The Seattle, Washington, environmental blog *Grist*, which had run the first posting "to implicate industrial hog farms," by Tom Philpott, posted a later piece headlined "Don't jump to conclusions on swine flu and pork production."[26]

Yet the bloggers were already at work, gathering and transmitting information. Salon.com published a piece by Andrew Leonard (September 3, 2009) referring readers to the "definitive article . . . on the intersection between swine flu and concentrated animal feeding operations (CAFOs)" (Leonard 2009; see also Brainard 2009). This article, published by Charles W. Schmidt two days earlier in *Environmental Health Perspectives*, the journal of the National Institute of Environmental Health Sciences, offered an evocative metaphor for the unique accelerating role played by CAFOs in a flu epidemic: "The best surrogates [for CAFOs] we can find in the human population are prisons, military bases, ships, or schools. But respiratory viruses can run quickly through these [human] populations and then burn out, whereas in CAFOs—which often have continued introductions of [unexposed] animals—there's a much greater potential for the viruses to spread and become epidemic" (Schmidt 2009, A395–A396).[27]

By 2009 it should have been old news that concentrated confinement (requiring prophylactic antimicrobial therapies) plus species proximity produced interspecies transmission. A 2007 study in *Environmental Health Protection* made that unambiguously clear:

> The industrialization of livestock production and the widespread use of nontherapeutic antimicrobial growth promotants has intensified the risk for the emergence of new, more virulent, or more resistant microorganisms. These have reduced the effectiveness of several classes of antibiotics for treating infections in humans and livestock. Recent outbreaks of virulent strains of influenza have arisen from swine and poultry raised in close proximity. . . . Concern about the risk of an influenza pandemic leads us to recommend that regulations be promulgated *to restrict the co-location of swine and poultry concentrated animal feeding operations (CAFOs) on the same site* and to set appropriate separation distances.   (Gilchrist et al. 2007; my italics)

Not surprisingly, when speculations that this new epidemic resulted from CAFOs rapidly emerged in the blogosphere, they generated the inevitable push-back by industry, which employed the classic technique of trivialization, associating the accusation with children's cartoons. "This is not the Porky Pig Plague," advised Robert Roy Britt on the Web site LiveScience.com two days before the World Health Organization announced that it would stop using the swine flu moniker, preferring instead the scientific term for the disease. "'Rather than calling this swine flu . . . we're going to stick with the technical scientific name H1N1 influenza A,' said WHO spokesman Dick Thompson" (Britt 2009).

Britt's move to distance the threat of swine flu from popular family cartoon icon Porky Pig—and thus from the image of pork as a wholesome family food—had limited success, as evidenced by the downturn in pork prices caused by the swine flu epidemic. Pork producers lost 5.3 billion dollars since the beginning of the swine flu crisis, *USA Today* reporter Paul Davidson reported in November 2009. And government attention to one facet of the crisis—the lag in swine flu vaccine production—didn't address the economic aspect. "The U.S. government is rushing to deliver swine flu vaccine at record speed, but there won't be a quick fix for pork producers, who have been financially battered by the virus. Unfounded fears that the H1N1 virus can be contracted by eating pork have contributed to the worst financial crisis in the history of hog farming" (Davidson 2009).

Despite the industry's attempt to obscure the connections between H5N1 and H1N1 (or between Chicken-licken and Porky Pig), the practice of high-density industrial animal farming may have been implicated not only in the accelerated outbreak and spread of the disease but in the hold-up in medical response to the threat as well. As early as April 2009, with the earliest doses at least six months away from production, the U.S. Centers for Disease Control warned that there would probably be a swine flu vaccine shortage (DeNoon 2009). By the following October, that projected shortage had become a reality. In early November, the Centers for Disease Control published a "Q & A Regarding 2009 H1N1 Influenza Vaccine Supply." The Web publication attributed the vaccine shortages to several factors: the tight timeline required of manufacturers, the failure of the virus to grow at the rate required to produce a large quantity for vaccine manufacture, problems with the potency of the virus grown in eggs, and the government's choice to ship small quantities of virus rather than waiting for the time when a complete shipment would be available. Yet the CDC omitted to list another problem long known to exist. The shortage of fertile eggs has been a continued problem to the vaccine industry. As Karl G. Nicholson reported in *Expert Review-Vaccines*, "in the event of a pandemic, the necessary number of high-quality fertile hen's eggs . . . is unlikely to match global needs" (Nicholson 2009).

An exchange between National Public Radio host Robert Siegel and Robert Belshe, director of the Center for Vaccine Development at the St. Louis

University School of Medicine, reported by Frank James on the NPR blog "the two-way," offered a more comprehensive explanation for the vaccine shortage:

> The old-fashioned method of making flu vaccine, which has been used for at least 50 years, is the method largely used around the world, requires millions of fertilized eggs into which the virus is injected, then grown.
>
>> ROBERT: Part of the process when the country orders up a bunch of vaccine is you've got to get these roosters and hens to work.
>
>> BELSHE: That's right. *These are not the ordinary eggs you buy in the grocery store which are generally not fertilized.* But it requires a fertilized hen's egg.
>>
>> So you've got to have a rooster and a hen get together, do their thing, then the eggs have to be laid. And then they have to be incubated until 10 days old for the idea[l] culture medium. . . .
>
> Not only is the egg method time-consuming but it is subject to the vagaries of agriculture. Also, if a pandemic flu of the avian variety arrives on the scene, the chickens who lay all those eggs could be susceptible, further slowing production.   (James 2009; my italics)

Yet even here the discussion is needlessly vague: what *are* the "vagaries of agriculture" referred to by NPR's Frank James?

The vaccine production problem is more than a simple supply-and-demand problem: it also consists in the difficulty of obtaining certifiably healthy chickens to lay the eggs that will protect human beings from a swine-to-human transmitted influenza. On September 15, 2009, four corporations were approved by the Food and Drug Administration to manufacture the H1N1 vaccine: MedImmune LLC (a wholly owned subsidiary of AstraZeneca with headquarters in Gaithersburg, Maryland); CSL Limited (in Australian corporation); sanofi pasteur Inc. (a division of sanofi-adventis group, headquartered in Lyon, France); and the Swiss giant, Novartis Vaccines and Diagnostics Limited (FDA 2009). While testimony of the executive vice president for operations of MedImmune LLC before the House Energy and Commerce Subcommittee in mid-November gives a good flavor of the complexities of vaccine manufacture, for our purposes the most interesting aspect is the regulations governing the production of those all important eggs in which the virus must grow.[28]

Eggs to be used for vaccine production must be laid by hens that are SPF, or "specific pathogen free."[29] Since the vaccine manufacturers are all multinational corporations, concern about pathogens in chicken flocks is global as well. A report filed in 2006 by Biosecurity Australia, "Contingency Import Policy for Specific Pathogen Free (SPF) Chicken Eggs," further clarifies what the "vagaries of agriculture" actually mean in the case of vaccine production: "an interruption to domestic supplies of SPF eggs, resulting from a disease 'break' in one or more of the then existing SPF flocks could have serious consequences for this country. . . . Recently, one of the two SPF flocks maintained by the sole Australian supplier became infected with chicken anaemia virus. If there is a disease outbreak in the other flock, the only option to ensure continued access to address these essential needs will be to import SPF eggs" (Biosecurity Australia 2006, 2).[30]

Why are these SPF flocks so precious and so rare? I have suggested that the slowdown in egg supply, the shortage of sterile chicken eggs for virus incubation, and the resulting slowdown in vaccination production can be traced back to the greater risk of infection linked to high-density confinement practices. But there is one additional factor in the scarcity of eggs, of course: the elimination of the fertile egg from ordinary market egg production.[31] Because the sex of day-old birds is very difficult to determine, specialized (frequently Japanese) chick sexers are employed by the poultry industry to distinguish male from female chicks. Using a technique of close examination of the chick's cloaca developed by Japanese scientists in 1924, these skilled employees function as a crucial cost-management point for the industry (Percy 2006, 47; see also Percy 2002, 58). While female chicks grow up to become valuable layers for the egg industry, their eggs are sterile, by design: male chicks are largely viewed as without value by the industry and, once sexed, are killed or even discarded alive within a week of hatching.[32] There is an ironic circularity to the story that the distribution of the swine flu vaccine to the American public may have been held up because this fifty-year-old method of vaccine production relies on a culture medium—the fertilized hen's egg—that has been made scarce due to the very nature of industrial farm production.

I began by comparing fear over avian flu to the story of Chicken-licken, whose alarm at that unexpected acorn fomented a riot, creating an opening for Fox-lox to engage in his own form of shock doctrine. I have been suggesting that any real conception of safety must transcend the Western emphasis on technical surveillance (and scientific production) to take into account modes of experience and forms of production beyond the metrics of expert science. Like Chicken-licken, we may find that the best counsel comes not from media pundits or science experts but from the forms of folk wisdom and tacit knowledge still embodied in the older form of family poultry raising. For despite the prominence of the Western industrialized model of poultry farming with its promised "biosecurity," the global dominance of that model of poultry farming is far from complete.

Although poultry growers in Thailand have successfully transformed the poultry industry under the guidance of Dhanin Chearavanont, whose giant factory farms and contract growers have effectively marginalized the traditional method of backyard chicken farming, other Southeast Asian nations have taken a different route (Greger 2006, 214–215). In Laos and Cambodia, an alternative model dominates, in which many small farmers grow small numbers of unconfined chickens, and the same model is followed in Africa and South America. "Despite efforts to develop intensive poultry production, family poultry (FP) are still very important in developing countries," explains E. F. Guèye. "In most developing countries, the keeping of poultry by local communities has been practiced for many generations. FP keeping is a widely practiced activity. More than 90% of rural families in most developing countries keep one or more poultry species (i.e., chickens, ducks, guinea fowls, geese, pigeons, etc.) and all ethnic groups tend to be involved in FP production" (Guèye 2005, 37). These "family poultry management systems" have become increasingly significant models for so-called developing countries both because they adapt a system of chicken raising that has been working successfully for many generations and because the distinctly gendered aspects of family poultry markets (in the third world, poultry keepers still tend to be mostly women) provide valuable economic autonomy and agency to rural women (Guèye 2005, 39). Like capitalism itself, then, industrial poultry farming is a permeable institution, both surrounded and infiltrated by those small family poultry systems (Gibson-Graham 2006).

Let's return to plumb the irony of the industrially produced circle we have charted, from avian flu to swine flu and back to the threat of avian flu. It is worth remembering that the family poultry management model is still very much a part of the global reality of poultry farming and has not yet been replaced entirely by industrial poultry production on a massive scale. We should remember too that this modern poultry industry is one among many industrial farm animal production systems. Fertile eggs from disease-free small flocks are not a thing of the past. The very permeability of the industrial farm animal production system is good news, just as to Gibson-Graham the evidence of capitalism's fissured nature opened up new possibilities for reinventing an institution from within.

What might this mean for the tale of Chicken-licken, and for the dream of bird flu with which I began? To me it suggests that conceptions of risk and safety must take the whole circle of farm production into consideration. It must extend beyond the farmed animals to the farmers and other workers engaged in their care and processing, and beyond technical surveillance to an awareness of the relationships not only between animals but also between animals and human beings, that exceed in complexity any simply technical risk management strategy.

Fellow-Feeling

"No sentiment or fine feeling enters into the life of the modern hen in an intensive poultry farm. Hatched early in the year, she is reared with one object, to lay eggs before the winter sets in. Once she lays, systematic feeding and exercise keep her at work."

–Anon., The Back Garden Hen

I had been raising chickens for a while—sitting with them in the morning as they scratched for insects in the dirt, making sure they were safely on the perch when I closed them in at night—when I began to think about fellow-feeling. The curious word came to mind when I thought about my feeling for my chickens. A kind of empathy—born in the intimate encounters that are so much a part of chicken farming: pouring the birds scratch grain and clean water as they mill noisily around my feet; enjoying the variety of their sounds, from soft clucks of contentment to urgent churrings when they turned up a worm; grabbing them by the legs and swinging them off the perch at night and holding them squawking upside down, their wings flapping, so I could dust pyrethrum (a mite remedy) on their butt feathers and under their wings; watching the hens lift and fluff their wing feathers to shelter their chicks. Fellow-feeling: the sense that my chickens are fellow creatures.

Chickens as fellow creatures? There was no question about that to Miss Nancy Luce, who also raised chickens more than a hundred years ago on the little island of Martha's Vineyard, Massachusetts. I had come across Miss Nancy in the archives of the university where I teach. Photographs of her with her chickens jostled with a pamphlet she wrote and a few short papers about her in a dusty manila folder in the special collections room. Small-scale farmer and poet, she raised chickens in backyard in West Tisbury, Massachusetts, during the 1860s and 1870s. To the tourists who visited her in hired livery carriages, drawn to her as a local curiosity, she sold hens' eggs and the privately printed pamphlets of her own poems. An article filed with her papers told a grim tale: "Her life a village tragedy, her publishing office an old leather trunk, her confidants hens, and her poetry unique, poor Nancy Luce lived out her sixty-nine years of poverty and ill-health in a humble farm-house on Tisbury Plain (Martha's Vineyard), and for forty-nine of those years, after her parents died, without human companion" (Clough 2006). But for me what stood out was the companionship Nancy Luce shared with her chickens. Although she raised them for their eggs, they gave her something else beyond that simple transaction. Something perhaps unquantifiable, something in the zone of fellow-feeling.

That something else that I puzzled over in Nancy Luce's story has puzzled economists, too, ever since the writings of Adam Smith. The "father of economics" was also interested in the idea of fellow-feeling. It featured prominently in his treatise *The Moral Sentiments* (1759), where he explored the workings of this complex emotion, central (he argued) to a fully functioning human community (Smith [1759] 1976). Yet in his foundational articulation of modern economic theory, *An Inquiry into the Nature and Causes of the Wealth of Nations* (1776), written only seventeen years later, that focus on emotional complexity is replaced by the development of a notion of economic self-interest that would shape Western culture for centuries (Smith [1776] 1904). Smith put it vividly in book I, chapter II: "It is not from the benevolence of the butcher, the brewer, or the baker, that we expect our dinner, but from their regard to their own interest. We address ourselves, not to their humanity but to their self-love, and never talk to them of our own necessities but of their advantages" (Smith 1904, I.2.2).

From benevolence to self-interest, from humanity to self-love: rural sociologists and agricultural historians such as Deborah Fitzgerald and Stephen Striffler have vividly traced the agricultural journey from the small flocks belonging to farm women such as Nancy Luce to the giant poultry farms where distant shareholders rely on armies of low-wage, often immigrant workers to produce profit through vertically linked corporations. It is a familiar story by now: the transformation of chicken farming from a casual, undocumented part of the small farm economy to the largest contemporary agricultural industry whose giant high-biosecurity grow-out buildings, each holding twenty thousand birds, stretch across the landscape of the southern United States.

Yet there are indications that times are changing in the chicken business. Large-scale poultry corporations such as Tyson and Cobb-Vantress seem to be feeling the need to reframe their massive operations to appeal to consumers interested in rediscovering the small-scale farming model. In June 2007 the *New York Times* reported that Tyson Foods had committed to converting twenty of their processing units to the production of antibiotic-free chickens to be sold fresh to customers.[1] In 2010 Cobb-Vantress Inc. introduced its "Cobb-Sasso, a "colored chicken, bred from a 'rustic' brown female mated with a white male" for "robust health and well-being, ideally suited to free range and organic systems, as well as less intensive indoor production."[2] The Cobb-Sasso sales brochures feature a photograph of white and parti-colored birds roosting in low tree branches as well as on a roost in what looks like a high-density indoor poultry house.

This return to the farmyard may have been spurred on by agricultural writers such as Wendell Berry, Barbara Kingsolver, Joel Salatin, and Michael Pollan (Berry 1996; Kingsolver 2007; Salatin 1993; and Pollan 2006). Or perhaps—as the *Times* recently speculated—it was the other way around: "Many of these writers say they are responding to the increased public appetite for food's back story. As they reveal their personalities, histories, and insights, they bridge the distance between the people who grow food and the people who eat it" (Bowen 2007). Today's "farmers who write" not only remind me of farmer-poets such as long-ago Nancy Luce; they may be reminding consumers that there is another way of raising our food. Many of the writers own or operate farms themselves—small ones, or even medium-sized ones—and sell their goods "directly to the public, either through farmers' markets or community-supported agriculture programs, or C.S.A.s [Community Supported Agriculture organizations] in which customers purchase shares of a farm's harvest" (Bowen 2007) Taggart Siegel's award-winning documentary film, *The Real Dirt on Farmer John* (2005) explores the emotional implications of this shift to community-based farming, showing the strong connection forged between Angelic Organics, "Farmer John" Peterson's Chicago-area CSA, and its twelve hundred shareholder families. As one of the farm workers observes, "people come here to remember what some of the basic things in life are: where your food comes from, your neighbors, who your farmers are." (Ebert 2006; see also Siegel 2005).

Feeding and watering my chickens in the early morning sun I found myself wondering what chicken-farmer-writers would say about the back story of chicken raising. And I wondered what Adam Smith would make of this changing moment in agricultural production. How would he explain the changed relation these writers are documenting to the crops we farm (animal as well as vegetable) and the changes in the structure of our agricultural economy? Could there be an economic meaning to fellow-feeling? I thought of three women involved in chicken farming between the end of the nineteenth and the beginning of the twenty-first centuries—Nancy Luce, Betty MacDonald, and Linda Lord. Two of these women could be described as "farmers who write," and while the third seems to lack the leisure time for writing, a skillful interviewer gave vivid voice to her contemplative powers. What role did fellow-feeling play in their experiences of the agriculture of their day?

As I would go on to discover, the changing pattern of their experiences of fellow-feeling was a disheartening one. While Nancy Luce, as a small-scale chicken farmer, seemed to mingle fellow-feeling and economic good sense without any difficulty, my mid-century chicken farmer Betty MacDonald found herself alienated from her chicks and chickens as she struggled to manage the unwieldy scale and schedule of her intensive egg farm. And Linda Lord—a low-wage worker in a poultry processing plant—displayed no fellow-feeling at all for the chickens she was killing, and precious little for her fellow workers on the factory line. If fellow-feeling led to community, and thus to a successful economy (as Adam Smith suggested several centuries ago), I thought, my survey suggested we need reinvent our notion of the economy and of fellow-feeling too, if we are ever to reinvigorate our communities. And perhaps that is precisely what we can be learning from those contemporary "farmers who write."

We will turn to Nancy Luce, Betty MacDonald, and Linda Lord in a moment, but first let's consider what fellow-feeling meant at the time of the birth of Western economics. In his classic study *The Theory of the Moral Sentiments*, Adam Smith argued that fellow-feeling was an essential ingredient in social interactions, providing the social glue that made all other aspects of society, and especially the economy, possible. Smith saw this sentiment as foundational to the accumulation of wealth because it induced the commitment to the human collective that was the prior condition of any economic transactions. As he saw it, fellow-feeling produced empathy and sympathy particularly with those less fortunate than we are: "This is the source of our fellow-feeling for the misery of others, that it is by changing places in fancy with the sufferer, that we come either to conceive or to be affected by what he feels" (Smith 1976, I.I.3).

Scholars have offered some trenchant critiques of Smith's proposition that the moral sentiment of sympathy creates beneficent social cohesion. Christopher Castiglia in particular, in a study of nineteenth-century American abolitionist reformers, challenges those who view sympathy as a "fellow feeling that prompts the privileged to imagine themselves in the place of the less fortunate" (Castiglia 2002, 36). He argues instead that by stressing affect, Smith's *Theory of Moral Sentiments* participates in the modern turn toward the regulation of interior life as the site of surveillance and normalization moves from the realm of governmental and civic life to the minds and bodies of individuals. In order to have one's suffering acknowledged, so the argument goes, one must stage it appropriately, so that it neither repels, angers, or bewilders the viewer. Emotional modulation is exacted as the price of social recognition: as Smith puts it, the suffering individual often ends up "by lowering his passion to that pitch, in which the spectators are capable of going along with him. He must flatten . . . the sharpness of its natural tone, in order to reduce it to harmony and concord with the emotions of those who are about him" (ibid.). Castiglia argues that in Smith's theory, this emphasis on internal emotions produces not connection but detachment, not egalitarian empathy but hierarchized regularization based on the distanced viewing of suffering as spectacle. In his critique, the benefactor's or reformer's experience of sympathy is actually an appropriation—and normalization—of the experience of the suffering Other, enabling through the process of shaping another's expression of feelings the perpetuation of precisely those structural inequities (of race, class, or gender) that initially triggered the viewer's sympathy. "Sympathy affectively naturalizes social hierarchy without necessitating government involvement, order being maintained as the result of a newly privatized internal civility" (ibid., 37).

While this critque spotlights a quality of Smith's work that I will touch on later—its capacity to restrict access to power even while posing to extend it—there is an embodied aspect to fellow-feeling that I want to explore first because it offers a key to the aspect of this moral sentiment that interests me now: its species-specific nature. While Castiglia argues that "Smith's formulation of sympathetic difference requires the *absence* of bodies," a close reading of Smith's text reveals that fellow-feeling induced a kind of somatosensory mirroring of the suffering of another person that is entirely reliant on a specific embodiment (ibid.).

We have a visceral reaction to the pain of others; we register it in our nerves and empathize with them based on our sense of sharing the same painful feelings in the same body parts. "Persons of delicate fibres and a weak constitution of body" may feel "an itching or uneasy sensation in the correspondent part of their own bodies" when they see "the sores and ulcers which are exposed by beggars in the streets." The sentiment is frequently targeted to specific bodily regions: "The horror which they conceive at the misery of those wretches affects that particular part in themselves more than any other; because that horror arises from conceiving what they themselves would suffer . . . if that particular part in themselves was actually affected in the same miserable manner" (Adam Smith 1976, I.1.3). Significantly, Smith's examples cross a number of social categories and incorporate a broad range of humanity in the deeply embodied experience of fellow-feeling, which in his recounting may induce empathy even with people outside the customary circle of gentlemen, such as a scabrous street beggar or, as he puts it later in his analysis, a woman in childbirth or a man suffering from insanity.

The one boundary fellow-feeling could not cross, in Adam Smith's model, seems to have been that of species. Fellow-feeling in Smith's terms seems premised on the awareness of bodily similarity. An injured leg speaks to a leg that might be injured; skin that is suppurating or inflamed resonates to skin that is still smooth. And while Smith himself limits the amount of fellow-feeling that a man may experience for a woman in the pangs of childbirth, his "theory of sociality" incorporates fellow-feeling as a support for "the 'two great purposes of nature': 'the support of the individual, and the propagation of the species'" (Sugden 2002, 84). We need social organization to create the security and material goods necessary for human populations to grow and thrive. Because species propagation requires gathering in societies and producing wealth, membership in the human species constitutes a boundary-marker. There is in Adam Smith no suggestion that the propagation of other species not as wealth but as objects of our fellow-feeling can have any survival significance. How Adam Smith would have pondered the story of Miss Nancy Luce.

Miss Nancy Luce

When she published her book of poems and advice for chicken doctoring in 1875, Miss Nancy Luce wrote at the dawn of the U.S. poultry industry. The U.S. Department of Agriculture had been signed into being only thirteen years earlier, the first poultry magazine in the nation (*The Poultry Bulletin*) had begun publication only five years earlier, the American Poultry Association had been organized only two years before, and it would still take another eleven years

FIG. 15.
"Miss Nancy Luce and Pinky."

Reproduced with the permission of Rare Books and Manuscripts, Special Collections Library, the Pennsylvania State University Libraries.

FIG. 16. "Miss Nancy Luce and Hens."
Reproduced with the permission of Rare Books and Manuscripts,
Special Collections Library, the Pennsylvania State University Libraries.

for the Hatch Act to create the agricultural extension agencies so powerful in defining so-called modern chicken farming (Hanke, Skinner, and Florea 1974, 35, 36, 52). In its early days, agricultural extension education framed chicken farming as a project particularly suited to women. The first speaker at a Farmer's Institute to address the topic of poultry raising was Mrs. Ida Tilson, of West Salem, Wisconsin, who spoke of her success raising chickens (65). The newly created extension agents appealed to women's long history raising chickens when they targeted the farm wife, who "above all . . . wanted her flock to contribute to the family income. She had confidence in its ability to do so. With few exceptions, the farmer himself took the opposite view, offering resistance rather than wholehearted acceptance" (66). The first poultry inventory in 1840 estimated that there were 98,984,232 head of poultry being raised in the United States, and women dominated poultry production in the nation until the end of World War II, when "large-scale egg and broiler production [had] gradually pushed women out of the poultry industry" (Sachs 1996, 107).

From her vantage point at the beginning of this process of agricultural transformation, Miss Nancy Luce wrote of her chickens as companions in the struggle for survival rather than as products. Her poems—many of them addressed to her chickens—speak vividly of her fellow-feeling for the suffering of others, both animals and humans, as does the veterinary advice pamphlet that she packaged as an appendix to her volume of poetry. Consider, for example, these verses to her favorite hen, Pinky.

POOR LITTLE HEART.

Poor Tweedle, Tedel, Bebbee, Pinky. She is gone. She died June 19th, 1871, at quarter past 7 o'clock in the evening, with my hands around her, aged 4 years. I never can see Poor little dear again.

Poor Pinky, that dear little heart,
She is gone, sore broke in her,
Died in distress, Poor little heart,
O it was heart rending.

O sick I do feel ever since,
I am left broken hearted,
She was my own heart within me,
She had more than common wit.

Poor Pinky's wit, and she loved me so well,
Them was the reasons,
I set so much by her,
And I raised her in my lap too.

. . .

I hope I shall never have a hen, to set so much by again,
From over the sea, she was brought to me, one week old,
I raised her in my lap,
She loved me dreadful dearly.

She would jam close to me,
Every chance she could get,
And talk to me, and want to get in my lap,
And set down close.

And when she was out from me,
If I only spoke her name,
She would be sure to run to me quick,
Without wanting anything to eat.

She was sick and died very sudden,
Only two hours and a quarter,
About fifteen minutes dying.
Bloody water pouring out her mouth,
And her breath agoing, Poor little heart.

O dreadful melancholy I do feel for my dear,
She laid eggs till three days before her death,
She laid the most eggs, this four years around,
Than any hen I have on earth.

. . .

Consider how distressing sickness is to undergo,
And how distressing in many ways,
My parents' sickness, a number of years,
Caused them to sell cows, oxen, horses, and sheep,
English meadow, clear land, and wood land,
Consider how distressing sickness is in many ways.   (Luce 1875)

The fellow-feeling expressed by Miss Lucy's heart-felt poetic narratives of her hens' illnesses is echoed in the appendix, "Doctoring Hens," with its urgently expressed medical (or as we would now describe it, veterinary) advice: "Human, do understand how to raise up sick hens to health. Some folks do not know how to doctor hens, they doctor them wrong, it hurts them, and it is dreadful cruel to let them die. It is as distressing to dumb creatures to undergo sickness, and death, as it is for human, and as distressing to be crueled, and as distressing to suffer. God requires human[s] to take good care of dumb creatures and be kind to them, or not keep any" (Luce 1875).

Her medical advice is specific and pragmatic. For a blocked cloaca she makes a no-nonsense appeal to womanly resourcefulness: "If a hen goes on her nest, and try to lay an egg, and cannot, and there most all day, then a skin of an egg is in her, she will certainly die if the skin of egg is not took out of her; someone has a small finger, and common sense, take the skin of egg out of her, then she is all right. I cure them so." For an intestinal obstruction she recommends massaging the hen's stomach, giving a dosage of Epsom salts gauged by the smell of the hen's breath, and then hand-feeding. "Folks bring hens to me in this disease, to the point of death, been sick a long time, I cure them in five days . . . do as I tell you and you can cure them" (Luce 1875).

Miss Nancy Luce clearly felt empathy and concern for the birds who were her companions. The final verse of the poem "To Pinkie" makes that clear when it asserts the similarities between animal and human suffering, just as her advice manual compares veterinary and human medical ailments and treatments. There seems to be little boundary between medical and veterinary matters, or between the hen's anatomy and her own. Nancy Luce suffered from illness much of her adult life; without much of a leap we could imagine the blocked oviduct or the intestinal obstruction as her own. Nor does she see any gap between sentiment and economic practicality: for a chicken farmer, tender care will be profitable. As she explains, "I bought a young hen last year, she was dreadful wild, and when one week was at an end she came to me, and let me take her up, she keep still, and eat out of my hand, she remains gentle ever since, and a good hen to lay eggs" (Luce 1875). Refusing the philosophy that links the right to care with the capacity to articulate feeling, she asserts that the simple practice of keeping animals entails the obligation to provide them with humane treatment. Fellow-feeling is central to Nancy Luce's psychic economy: it is how the energy she devotes to the relationship with her chickens, and with the broader human society to which she sells their eggs, is expended, regulated, and replenished.[3]

Betty MacDonald

As a child of the 1950s, I loved MacDonald for her wonderful Mrs. Piggle-Wiggle series of children's books. As an adult raising chickens, I also love her for *The Egg and I* (1945), the hilarious account of her struggles as a chicken farmer on the Olympic Peninsula of Washington just after World War II. At the top of the U.S. nonfiction lists within a year of its publication in 1945, this memoir captures a woman's experience of chicken farming at the conclusion of the agricultural transformation whose early glimmerings Miss Nancy Luce documented.

Chickens and eggs had an ever-growing consumer market during and after World War II, for they were the only products of animal agriculture not subject to wartime rationing. By 1945 chicken farming had been reinvented as a task for men, particularly for war veterans hoping to establish a profitable place in the postwar economy. Poultry keeping took on a systematic, scientific cast, as new methods of flock management and feed ratio aimed at producing a larger product more quickly and economically. This is the world entered by Betty MacDonald and her husband, Bob, an ex-Marine war veteran, once they decide to take the plunge into chicken rearing.

For the MacDonalds, fellow-feeling for chickens is definitely the man's emotion. When, on their honeymoon, Bob begins to talk about his experience raising hens and his interest in returning to chicken farming, his feeling for chickens combines ardor and economics. As Betty tells it, Bob speaks of his job as a supervisor on a large chicken ranch "with the loving care usually associated with first baby shoes. When he reached the figures—the cost per hen per egg, the cost per dozen eggs, the relative merits of outdoor runs, the square footage required per hen—he recalled them with so much nostalgia that listening to him impartially was like trying to swim at the edge of a whirlpool" (MacDonald 1945, 38).

The humor in *The Egg and I* lies in Betty's sentimental education as a wife-turned-chicken farmer. While Bob's feelings for his chickens are tender, Betty's take a different turn. Before married life, her only contact with chickens had been watching Layette, her grandmother's Barred Rock hen, lead her brood of "fourteen home-hatched fluffy yellow chicks through the drifting apple blossoms." As her farming experience increases, her feelings ripen and twist, until she has "Learn[ed] to Hate Even Baby Chickens" (138). The daily farm grind vanquishes the memory of Layette: "This sentimental fragment of my childhood was a far cry from the hundreds and hundreds of yellowish white, yeeping, smelly little nuisances which made my life a nightmare in the spring," she explains wryly (138).

Many factors intertwine to ensure that Betty will hate the chickens she and Bob raise on their wilderness chicken ranch. MacDonald's memoir milks the humorous meaning from each of them: her experience as a dutiful postwar wife who subordinates her goals to those of her husband despite being an aspiring writer; the travails of new motherhood in their isolated rural setting; her lonely life as a farm wife and mother far from a supportive community. But the stand-out factors are the scale and gender of modern chicken farming. Unlike Nancy Luce, the MacDonalds are enmeshed in modern intensive poultry farming, in which high-volume egg production requires efficient feeding and watering, careful record keeping, and the rigorous culling of unproductive layers. And throughout all of this, the farmer at hand is implicitly, indeed explicitly, gendered male. Bob is "the best chicken farmer in our community" because "[he] was scientific, he was thorough, and he wasn't hampered by a lot of traditions or old wives' tales." Like other modern poultrymen, he separates "breeding and egg raising," understanding them as "two separate industries." He studies egg prices and calculates the ratio of input (feed) to output (eggs). And he too has opinions about what makes a poultry farm succeed or fail: "the secret of success in the chicken business for one man was to keep the operation to a size that could be handled by one man" (146). Any farm wife who would work side-by-side with such a paragon must be, MacDonald quips, "part Percheron" (146).

The gendered transformation of poultry farming MacDonald represents with such humor also had serious implications, particularly for working-class women. Recall that, traditionally, the "egg money" had been held separate from the general farm income. It was the private property of the wife, to use as she wished. Thus, raising chickens gave farm wives not only the power to amend the farm diet when necessary but also precious economic (and therefore at times even social) autonomy. When the rise of extension education in poultry farming destabilized this arrangement, producing a new image of the scientific male farmer in the 1930s and 1940s, women lost control over an aspect of their daily lives and crucial income as well. Rather than enjoying the security of being able to provide for themselves, they were now subject to the vagaries of an economic system in which they no longer had a significant part. It is little wonder that they felt less fellow-feeling for their chickens, entrapped as they were in an economy that showed increasingly little concern for women's economic or social welfare. The impact of this economic trend on women, and on the broader communities they anchored, can be gauged in my final example: the experiences of poultry worker Linda Lord.

Linda Lord

Like Nancy Luce and Betty MacDonald, Linda Lord's story is intimately shaped by her surroundings. However rather than rural Martha's Vineyard with its occasional summer visitors, or the lush but lonely Olympic Peninsula, Lord's territory is hardscrabble Belfast, Maine. The chickens she encounters are not around her front door or in her parlor as were Nancy Luce's, or in whitewashed poultry houses adjacent to Betty MacDonald's wilderness cabin, but in a dimly lit factory at Penobscot Poultry where they came to be stunned, killed, scalded, plucked, singed, and packed for shipping.

The poultry industry had been a prominent part of the Maine economy since the time of Nancy Luce: producing mostly eggs in the 1860s but by the 1930s, when Betty MacDonald was getting into chickens on her northwest coast island, shipping "New York Dressed" chickens to out-of-state buyers (Chatterly and Rouverol 2000, 97). After World War II, poultry had become Maine's most important agricultural crop. Even when Maine's poultry industry consolidated into only a few large processing firms, Penobscot Poultry held on, remaining as one of only two Maine broiler processors (98). In 1971, about the time Lord started work as a poultry processor, Penobscot Poultry was severely damaged as an industry when it was found to have been dumping pollutants into Penobscot Bay, along with another company, for which they were both fined. Later in the decade the company's environmental problems continued, when polychlorinated biphenyls (PCBs) were found in the flesh of chickens from three Maine poultry plants, Penobscot among them, leading to the destruction of 1.25 million birds. The error was finally attributed to one flawed batch from a Ralston Purina feed mill, rather than to Penobscot Poultry, but the damage was done (98). By the 1980s Maine's struggling economy was undergoing deindustrialization along with much of the rest of the Northeast, requiring the state to shift to a tourist economy. Penobscot Poultry finally closed in 1988. Historians hold differing opinions about the factors that contributed to the plant closing, a topic debated hotly all over Belfast, but clearly among the mix were financial problems, mismanagement, and the intensifying demands of vertical integration. Linda Lord, who had given the plant more than twenty years of her life, hired there right after high school, registered both the economic and the emotional effects of that transition, as Cedric N. Chatterley and Alicia J. Rouverol document in their moving photographic history, *"I Was Content and Not Content": The Story of Linda Lord and the Closing of Penobscot Poultry* (2000).

Lord held many positions at Penobscot Poultry during her two decades of work, jobs ranging from low-pay poultry transferring (putting the killed and plucked birds on shackles for the eviscerating line) to "top-pay" poultry sticker (killing the stunned birds by putting a knife through "the vein right in by the

FIG. 17. "Linda Lord at Work on Closing Day at Penobscot Poultry."
Used by permission of Cedric N. Chatterley.

jaw bone") in the room they called the "blood tunnel" or "hell hole." As she
explains in her extended interview with the authors, her experiences with
chickens focused on death, dismemberment, and disability, rather than the
breeding, laying, and hatching that I found in the stories of Nancy Luce and
Betty MacDonald. As I read her interview and gazed at the powerful photographs accompanying it, I wondered how to bring Linda Lord's very different
experience of chicken farming—factory farming, as it was for her—into conversation with the stories of the other women. Her interview gives poignant articulation to the complex psychic and material economies of industrial chicken
farming. Although (or more likely because) she hit the top of her personal pay
scale—at $5.69 an hour, same as the company's truck drivers—as a sticker in the
blood tunnel, Lord has a clear grasp of the economic realities of her form of
employment. "The way I look at it now, the chicken business here in the state
of Maine is just about phased right out. Because it's costing too much for us, for
the grain to be shipped. You have to pay electricity, you have to pay the fuel. You
know, it's a sad thing really, because it's put a lot of people right out of work"
(Chatterley and Rouverol 2000, 12, 34).

But what were her specific experiences of fellow-feeling? I wondered. She
acknowledged sympathy for her fellow poultry workers based on the physical
injuries they shared. Whether she was standing on the assembly line hanging
dead poultry, or working awash in blood grabbing stunned chickens off a moving overhead line, Lord explains in her interview, she was prey to the same

problems that characterized all of the people who "work on the lines": warts, tendonitis, and blood poisoning. There's an ailment specific to the industry, Lord explains, "You have like a rash break out all over you from the chickens . . . which eventually will blister right up with little pus sacks, and the skin will peel right off your hand, and they call it—'chicken poisoning' is what they call it" (5). Lord has suffered from blood poisoning, and she is blind in her right eye as a result of an accident on the assembly line, an injury for which she finally got a partial settlement after three years working with the workers' compensation lawyers at her union. It was no victory, though: as Lord points out. Even drawing on the brutal calculus of compensation, the disability money she drew from the company did not offset the economic cost of her injury when it barred her from other employment possibilities once the plant closed: "They had a way that if—you know, the loss of a finger was so much, and a loss of a hand was so much, or an arm was so much, or an eye was so much, and that's how it went. But I don't think they really realize just what one eye—how it limits you in a lot of things. And it hurts you for good jobs, too" (59).

As I read Lord's comments, I thought of Adam Smith's observation that fellow-feeling leads us to empathize with the ailments of others. Things seemed different for Lord. Responding to an interviewer's question, she made it clear that whatever connection she felt to her fellow workers based on shared physical vulnerability was tempered by economic realities: her straightforward need for this job and knowledge that the actions of fellow-workers may jeopardize that income. In the late-twentieth-century climate of scarce jobs for low wages, Lord found a factory job doing poultry processing appealing because "that was just about the only place that was hiring . . . And I wanted a job" (2). Although roughly a generation earlier Nancy Luce urged people to realize that "it is as distressing to dumb creatures to undergo sickness, and death, as it is for human[s]," Linda Lord's relations to chickens are instrumental rather than emotional, and are focused not on their generative capacities (the eggs they lay) but on their destruction. She attributes her skill as a chicken sticker to a lack of fellow-feeling for the birds themselves.

SC [INTERVIEWER]: You know, the first day we photographed the hell hole, we were a little taken aback. It's a pretty gruesome scene in there. How did you feel about it when you began to work in there?

LINDA LORD: I was the type of kid growing up that nothing bothered me—blood or anything like that. So when I signed up for that job—of course I'd been in there and I'd watched and I had tried some, you know, on my breaks and stuff. . . . So I knew what I was getting into, and it didn't bother me, and I preferred working by myself than working by someone that might cause trouble for you on the line.

SC: How could people cause trouble for you on the line?

LL: Oh, throwing stuff, you know, not doing their bird, but trying to blame it
      on you. . . . And if someone wasn't doing their job on a bird, they could
      trace it back.

LL: So you thought that maybe the disadvantages of working with other people
      were big enough that you rather would have stuck it out by yourself in that
      room.

LL: Yeah.   (Chatterley and Rouverol 2000, 7)

In the life and death of the Penobscot Poultry company we learn some of
the emotional and material effects of the transformation of backyard chicken
farming to large scale poultry production. In its compartmentalization of the
life of the chicken, so that chicks are incubated and hatched in one division,
laying hens are housed and eggs produced in another division, and broilers
grown, slaughtered, and processed in a third division, Penobscot Poultry embodies the rationalization of life itself. Similarly, Lord's experience as a poultry
worker demonstrates how the industrialization, consolidation, and accompanying deindustrialization of regional poultry production both reframed and nar-
rowed the meaning of community for agricultural workers as well as the
meaning of agricultural work for the community of Belfast. Reading her story
and musing angrily on the injustice of our contemporary economic paradigm, I
found myself wondering whether a return to Adam Smith's writings could shed
light on this economic and social transformation, and maybe even give some
hope of an alternative model.

Adam Smith

Economists have long struggled to connect Smith's *The Theory of Moral Sentiments* (and its explorations of sympathy, empathy, and fellow-feeling) with the
book he published only seventeen years later, *An Inquiry into the Nature and
Causes of the Wealth of Nations* (1776). Indeed, as they have tried to show that
Smith's ideas of empathy and sympathy are compatible with notions of rational
choice theory, neoliberal economists have painted a picture of economic relations reminiscent of those Linda Lord encountered at Penobscot Poultry. They
have argued that human beings are guided not by feelings but by rational preference, and that individuals' deeds are shaped by a rational assessment of the
utility of possible choices of action. Thus they have reinterpreted sentiments
such as fellow-feeling as something closer to choice, making it possible to
explain why people engage even in altruistic or sympathetic activities, although
on the surface they seem of little personal utility, by quantifying the choices of
an individual or even a community (Sugden 2002, 64).

More recently, critical economic sociologists and feminist economists
have issued a strong challenge to such a notion of rational choice economics.

Framing a critique of neoliberal economics that offers an alternative economic theory attentive to issues of social justice, economist Robert Sugden refuses the neoliberal reinterpretation of fellow-feeling as preference. Instead, he argues that Adam Smith's notion of fellow-feeling exceeds its representational capacities, modeling not merely utilitarian preferences but also the interaction between different human beings' mental states. Sugden reminds us of what Adam Smith meant by fellow-feeling: "one person's lively consciousness of some affective state of another person, where that consciousness itself has similar affective qualities" (71). While these similar qualities could consist in either the shared pain or joy that we may feel in response to another's experiences, the crucial point to Sugden is that whatever the emotion shared, the experience of fellow-feeling is itself a good phenomenon. It gives us the inherently positive experience of being linked to another's consciousness, sharing their affective experience, whether pleasurable or painful. (The converse is also true, Sugden observes: failure to share another's feelings can cause us distress, irritation, or even pain.)

As Sugden reinvigorates Smith's analysis, he explains that we are motivated to participate in society, and to bind ourselves to its demands and restrictions, because there is a pleasure in the fellow-feeling we obtain thereby. "Fellow-feeling is an essential part of the technology by which relational goods—that is, social relations that have subjective but non-instrumental value to the participants—are produced" (81). As Sugden shows, Smith's discussion of fellow-feeling in the context of moral development moves us beyond simple instrumentalism into the realm of the psychic economy, delineating the broader motivations for our participation in society.[4]

Feminist economists Julie Cameron and J. K. Gibson-Graham have gone even further than Sugden, broadening the economic realm to include previously marginalized groups and modes of transaction previously viewed as non-economic (Cameron and Gibson-Graham 2003, 153). Redefining the economy "as an open-ended discursive construct made up of multiple constituents," they suggest that we employ additional categories of economic analysis such as production and reproduction, caring and nurturing, gift relations, environmental stewardship and social cooperation, all to further the feminist goal of improving the representational adequacy of our economic models. They show how we can introduce what they call "alternative axes of differentiation within the economy"—that is, new models for explaining how difference is produced and maintained—to disrupt our conventional assumption that human activity is divided into economic and the non-economic. We can map the true diversity of economic relations and practice by including alternative forms of transaction (and calculation), different modes of laboring and paying for labor, and a range of different kinds of economic organizations with different relations to surplus labor. This exercise in representing alternative modes of economic analysis can help us see how we might move toward "an economy in which the inter-dependence of all who produce, appropriate, distribute and consume in society is acknowledged and built upon" (153).

This dense economic analysis may seem a long way away from the equally dense materiality of the farming experience, but it does enable us to tease out the multiple forms of fellow-feeling expressed in Nancy Luce's poem to her chicken "Pinky." She both praises her for her tender attention and remains aware of her impressive capacity for egg production:

> She would jam close to me,
> Every chance she could get,
> And talk to me, and want to get in my lap,
> And set down close.
>
> And when she was out from me,
> If I only spoke her name,
> She would be sure to run to me quick,
> Without wanting anything to eat.
>
> O dreadful melancholy I do feel for my dear,
> She laid eggs till three days before her death,
> She laid the most eggs, this four years around,
> Than any hen I have on earth.

Once we think of economics as simultaneously incorporating many con-tradictory notions of value and different kinds of transactions, we understand Miss Nancy Luce's egg business as just one of several forms of exchange in which she participates. These also include her care for her hens in return for the affection she receives in return, the sale of her pamphlets of poetry and medical advice, and her exchange of medical care for others' chickens in return for the good will and companionship she hopes (if often in vain) to receive from her human visitors. When we read her biographer's lament that "Nancy's house had been a recognized sight for 'summer visitors,' . . . many used to drive 'up-island' to scoff, and some remained to buy eggs and poems. Others plagued Nancy, baiting her to see what she would do or say," we are able to link the via-bility of different forms of economy to each other (Clough 2006, 253). We understand that the illness and physical collapse of her mother and father led to the failure and sale of their farm, crops, and woodland, which led to Nancy Luce's dependence on the chickens not only for livelihood but also for the only fellow-feeling left to her. From the fellow-feeling for the chickens comes an awareness of the loss of fellow-feeling that she and her parents suffered due to their impoverishment and isolation.

Attention to empowerment and environmentalism, the issues intro-duced by feminist economists Cameron and Gibson-Graham, enables us to see more clearly the complex costs of modern chicken farming for Betty MacDonald and Linda Lord. We understand how Betty MacDonald's position as helping hand on the chicken farm managed by her husband reinforced her already-entrenched sense of disempowerment as a woman while her ability to tolerate the chaos and hard work provided the theme for her best-selling memoir. And we can see how Linda Lord's poultry processing work both empowered and disempowered her: how she took pride in being "the first woman in the industry to have a job in the sticking hole" and "the only one that stuck it out killing birds and stuff," and how she felt grief at being disabled and losing her employment (Chatterley and Rouveral 2000, 71). Moreover, we realize that both Lord and her community of Belfast suffered materially and socioculturally from the environmental disaster and economic collapse caused by a PCB-tainted feed mill, ultimately contributing to the closure of Penobscot Poultry.

Returning to Adam Smith's theories with the new perspectives introduced by feminist and critical economists, we can now understand the importance of conceptualizing the agricultural economy broadly enough to recognize its vital variety. As Cameron and Gibson-Graham have shown, such an economy has room—indeed must make room—for "multiple constituents" (Cameron and Gibson-Graham 2003, 153). As the stories of Miss Nancy Luce, Betty MacDonald, and Linda Lord reveal, the varieties of work in chicken farming, like the different sites in which the work takes place and the range of possible transactions that can result from it, will shape the extent to which any of us is able to experience fellow-feeling as foundational to community-building rather than merely consolidating existing structural inequities. For each agricultural model, like each kind of encounter with chickens, generates its own distinct capacity for fellow-feeling, and its own specific notion of community.

As I sit out with my chickens at the end of a long day, when I collect the handful of brown, green, blue, or white eggs in the early evening; when I augment my flock with a bantam hen and chicks bought at the poultry auction in Big Valley, where the Amish buggies jostle with the pickups and minivans; when I crate up some of my hens and cockerels to take to the Mennonite butcher in the next valley; and when I serve those birds to my friends in the form of a thick chicken corn chowder, I think that economists have gotten it wrong. In their focus on rationality, utility, and quantifiability, they have overlooked the possibility—present even in the work of their founder, Adam Smith—of creating wealth in its broadest sense, which includes the production of well-being through embodied connection to others, even other species. This possibility is embodied in models of backyard and sustainable chicken raising currently making a precarious place for themselves in the shadow of the modern factory farm. Through the pleasures and pains of their individual stories, Nancy Luce, Betty MacDonald, Linda Lord, like the farmer-writers of our contemporary moment, reveal that we have much to gain if we refuse the grim stricture: "No sentiment or fine feeling enters into the life of the modern hen in an intensive poultry farm."

Gender

"Why are women like chickens and chickens like women?"

–subRosa Art Collective, *Cultures of Eugenics*

Two versions of the same children's story have pride of place on my office shelf: a large-format illustrated pulp paperback version of *The Little Red Hen*, produced by the Saalfield Publishing Company in Akron, Ohio, in 1928, and a glossy child-sized hardback version of *The Little Red Hen* (Rand McNally Junior Elf Book) published in 1957. These children's books provide my point of entry to an exploration of the relations between women and the most liminal of livestock—the chicken. In the old children's folktale, Little Red Hen finds a wheat seed, asks the other farmyard creatures for help planting it, and when they refuse her, she goes on to plant, reap, thresh, and take the wheat to be ground at the mill. She mixes and bakes the bread, still without help, and in the end she eats the bread, ignoring the pleas of her fellow creatures to share it with them.

As a folktale, *The Little Red Hen* has generally been read as exhorting young children to work hard, accept responsibility, and share with others. Children's stories are more than simple didactic tracts, however. Feminist critics have shown that because such texts reflect their time and place, they frame what is considered gender-appropriate behavior at the time, even modeling that behavior to its young audience. But these folktales do even more than that, I want to suggest: they raise a number of linked questions about women and agriculture, and they do so through the figure of the chicken—or specifically, the little red hen.

Whether due to their widespread domestication, their small size, their relative ease of management, or their ability to subsist by scavenging, the chicken holds a potent liminal position between backyard and farm fields, between the "egg money" of the farm wife and the formal farm economy, between the private world of women and the public world of men, between the realms of animal agriculture and human reproductive medicine, and between the practices of traditional farming and the new world of industrial meat and egg production and *pharming* (breeding chickens genetically engineered to lay eggs that express pharmaceutical-grade proteins, used to create profitable new drugs).[1] I use the term *liminal* to refer to "those beings marginal to human life who hold rich potential for our ongoing biomedical negotiations with, and interventions in, the paradigmatic life crises: birth, growth, aging, and death" (Squier 2004, 9).

As liminal livestock, chickens play a central role in our gendered agricultural imaginary: the zone where we find "the speculative, propositional fabric of agricultural thought . . . which supplements the more strictly systematic, properly scientific, thought of [agriculture], its deductive strategies and empirical epistemologies" (Waldby 2000, 136).[2] Looking at chickens and chicken farming as they are explored in art of various kinds, we can uncover the basic unarticulated assumptions that help to shape the role of women in farming and the role of farming in women's lives. Because different artistic media offer us access to, and catalyze, different aspects of the agricultural imaginary, I turn in what follows to several children's stories, a novel, and a film to ask, adapting subRosa Art Collective's memorable question: "What does it mean, to feminism and to agriculture, that women are like chickens and chickens are like women?" (subRosa 2005).

Art and the Agricultural Imaginary

From the ancient children's fable *The Little Red Hen* to contemporary works of art in a number of different media, chickens are frequently represented as liminal livestock, and as such they articulate the complex intersection of women and agriculture. I will sketch out how this operates, beginning with analysis of the changes in the anthropomorphic character Little Red Hen that testify to the U.S. agricultural transition from family farming to intensive confinement agriculture. Ruth Ozeki's novel *My Year of Meats* further complicates the simple dyad of woman and chickens by exploring the connections between animal agriculture and human reproductive medicine and dramatizing the racialized nature of chicken consumption and distribution in the United States. John Fiege's documentary film *Mississippi Chicken*, an exploration of the lot of chicken processing workers in contemporary Mississippi, reveals the interlinked gender-, race-, and ethnicity-based oppression of women that are part of the contemporary poultry business. Another children's tale featuring a little red (or brown) hen, Katie Smith Milway's *One Hen: How One Small Loan Made a Big Difference*, suggests how even the well-intentioned efforts to improve the lot of one single mother in Ghana by giving her son the means to take up poultry farming ends up introducing not only new economic possibility but also the gendered structures of Western high-volume poultry farming into what was a very different agricultural mode. Taken together, these images offer no answers but rather suggest that the multiple practices, spaces, and forms of the woman–chicken relation embody some important questions that are integral to the topic of women and agriculture.

From Barnyard to Farmhouse: Poultry Moves Indoors

Consider two versions of *The Little Red Hen*: one published in 1928, only eight years after the passage of the Nineteenth Amendment gave women the right to vote, and the other published in 1957, one year after Bette Nesmith Graham invented "Liquid Paper," the so-called secretary's friend, a form of white paint "used to correct typing errors."[3] As its cover illustration suggests, the 1928 version of *The Little Red Hen* features a rather blowsy big hen—probably a Rhode Island Red, a highly popular hen in that era—whose capable aproned body conveys a sense of strength and responsibility extending beyond her many chicks to the farm itself. She is no vegetarian, believing that "fat, delicious worms . . . [were] absolutely necessary to the health of her children." In fact, when she finds the wheat seed, she is "so accustomed to bugs and worms that she supposed this to be some new and perhaps very delicious kind of meat."[4] Only after making inquiries does she discover that it is a wheat seed, which "if planted . . . would grow up, and when ripe could be made into flour and then into bread." She isn't much of an agriculturalist, either, being "so busy hunting food for herself and her family that, naturally, she thought that she ought not to take time to plant it." So she asks the Pig "upon whom time must hang heavily," the Cat, and the "great fat Rat with his idle hours" if they will help her by planting the wheat seed. "Not I," they all say. "Well then," says the Little Red Hen, "I will." And despite the cries of her chicks that she is neglecting them, her daily duties to find worms and feed her chicks, and the increasing fatigue of those "long summer days," she plants the wheat seed.

Come harvest time, the same thing happens. The hen's attention is "sorely divided between her duty to her children and her duty to the Wheat, for which she [feels] responsible," so she asks again "in a very hopeful tone . . . 'Who will thresh the wheat?'" Once again, the answer is a unanimous "Not I," and once again she sets to work. She threshes the wheat (alone) and drags it (alone) to the distant mill where she orders it "ground into beautiful white flour." Then the flour must be made into bread. Although far from a homebody and "not in the habit of making bread," still the Little Red Hen is determined and confident, knowing that "anyone can make it if he or she follows the recipe with care . . . [and] she [can] do it if necessary." So when once again the Cat, Rat, and Pig refuse to help, she bakes the wheat herself. At last, as she takes the aromatic loaves from the oven with an air of perfect calm, she suppresses "an impulse to dance and sing." "Then, probably because she had acquired the habit, the Red Hen call[s]: 'Who will eat the bread?'" However, when all the ani-

FIG. 18. "Little Red Hen circa 1928"

mals in the barnyard chorus "I will," she comes to her senses: "No, you won't. I will," she pronounces. And, the narrator concludes, "she did."

When we turn from the rich farming detail of the 1928 version to the 1957 version, which features a cuddly caricature of a Rhode Island Red hen, we can see just how much the cultural context for this children's story has shifted. This

FIG. 19. "Little Red Hen in 1957"

Red Hen sports black boots with silver buttons and a spotless white apron, and has a pair of very human front-facing eyes with prominent whites and a beak far more duck- than chicken-like. Unlike the earlier hen with her more customary brood of eight chicks, this somewhat Disney-like hen is pictured in the cover illustration with an appropriately small fifties' era family of three chicks. Not only the hen but the farm around her has changed. The illustrations include no insects for the hen to hunt for her family's dinner; no cat or rat

FIG. 20. "Little Red Hen at the Stove"

among the farmyard creatures to lazily refuse to help her with the wheat; no hard work or fatigue as she plants, reaps, and threshes the grain, or bakes the bread. Instead, the hen plants the "few grains of wheat" she has found "scattered about on the ground"; marvels at the "golden holiday clothes" the wheat seems to be wearing at reaping time; and when the Duck, Goose, and Fat Little Pig refuse to help her, finds a perfectly sized sickle, flail, and sack to reap the wheat, thresh the golden grains, and carry the sack on her back (along with a chick) to the mill.

Cheerful, tireless, tidy, and uncomplaining even when the Duck, Goose, and Fat Little Pig refuse to help her, this new model hen is also a natural in the kitchen, baking the bread herself without a moment of hesitation. But there the Little Red Hen's cheerfulness ends. "You remember that I planted the wheat and cut it, I threshed it and carried it to mill, I made the bread and baked it—and now all of you would help me eat it! No indeed!" she tells the Duck, Goose, and Fat Little Pig. As they lament their lot—"A little work and a little less sleep wouldn't have hurt me any"—she closes the farmhouse door on them firmly. The moral is clear: the Little Red Hen has labored not for herself but for her family. In the last scene of the 1957 version, we join the Little Red Hen inside the little red-roofed farm house where a framed portrait of the paterfamilias— a Little Red Rooster—hangs over the coal-fired stove. "But the Little Red Hen who had worked so hard was happy, and she sang as she cut thick slices of the fresh bread and spread them with butter and jam for her hungry chicks."

Drawn respectively from the dawn of women's suffrage and the postwar era of subservient secretaries and organization men, these two stories reveal a change both in the Little Red Hen and in the farm woman on whom she is anthropomorphically modeled. The Little Red Hen of 1928 is meat-eating, hunting, resourceful, and self-directed. While she has to negotiate the competing demands of her unappreciative chicks and her demanding, exhausting, working conditions—as we learn from the narrator—she does so with rueful pragmatism. She even drops her habit of asking for help—briefly losing her diminutive thereby, to become just "Red Hen"—so that by the story's end even her chicks seem marginal to her pleasure in claiming the delicious bread for herself. In contrast, the Little Red Hen of 1957 is a tireless worker, cheerful, and, above all, self-sacrificing. The story ends with her preparing to feed the freshly baked bread not to herself but to her hungry little chicks.

Livestock farming for meat is not addressed directly in *The Little Red Hen*. Possibly taboo for the increasingly urban, sheltered children who are the target audience, it appears only in an illustration in the 1928 version, where the pig is pictured with an apple in his mouth in allusion to his impending slaughter. Still these two versions of the classic folktale reflect a changing understanding of the farm and the farmer in twentieth-century America that was catalyzed by a transformation in chicken farming. As we know, chickens were not figured in as part of the farm economy but were raised almost as an afterthought by the farm wife; they ran freely in the farmyard and lived on scavenged grain and table scraps. The cockerels and pullets provided low-cost accessible protein for the family that helped them ride out the fluctuations in the farm market while the laying hens provided a product that could be bartered or sold for pin or egg money, giving a measure of economic autonomy to women otherwise without a private income.

Yet over the course of the twentieth century the face of U.S. farming changed dramatically. While more than one-third of the U.S. population lived on family farms in the early 1900s, by the 1990s that number had dropped to less than 2 percent, according to Lobao and Meyer (Lobao and Meyer 2001, 103). This decline was twofold, consisting not only in the precipitous drop-off in the number of people who attributed their family livelihood to farming but also in a dramatic change in the structure of agriculture itself. No longer small scale, single-family operations, farms were increasingly large-scale, vertically organized corporate ventures to which the farmer contributed labor, materials, and risk while the agribusiness retained the product and the largest share of the profits.

The role of women in this change in the nature of farming in the United States has been particularly hard to specify. Like the chickens they raised, which were not part of the livestock census nor figured as part of the farm economy, women have not counted in the Census of Agriculture, which "allows for only one self-defined operator per farm" (Lobao and Meyer 2001, 109–110). Like chickens, women too have been held apart from the public face of farming. As Berit Brandth has argued, "especially relevant to understanding gender relations in family farming is the separation of the household from the economy, the family from wider kinship groups, and the private from the public" (Brandth 2002, 110). This transformation in the place and personnel of farming took place on several scales, as Carolyn Sachs has shown. As chickens made the transition from marginal household creatures to livestock for the farm economy, they were increasingly raised not in small farm flocks but in the large-scale indoor barns of the poultry industry, where their control was increasingly the province of men rather than women (Sachs 1996).

Comparison of the two versions of *The Little Red Hen* reflects this transition-in-process. Since rationalized scientific agriculture has increased and the farm population has declined in the period between 1928 and 1957, the agricultural setting of the later version of the children's tale has become less realistic and more sanitized and fanciful. Certain details have disappeared that were important to the 1928 version, like the bugs and fat delicious worms that the hen views as "absolutely necessary to the health of her children," the ever-fattening pig whose time hangs heavy on his mind because of the impending end-of-summer slaughter; and the "great fat rat." Instead, by 1957 both the illustrations and the text reveal an antiseptic barn, silo, and windmill. And like the move to confined animal feeding operations catalyzed by the poultry industry, by 1957 the Little Red Hen and her chickens live indoors. These changes are all indicative of the role that chicken farming played in the consolidation of the new industrial agriculture, as rural sociologists, geographers, and historians have documented (Sawyer 1971; Sachs 1996; Bugos 1992; Striffler 2005; Midkiff 2004; and Horowitz 2006). In the contrast between the 1928 hen—an assertive hunter-gatherer mother who gathers worms for her chicks and lives on a very "real" and unsanitized farm—and the 1957 hen—a decorous farm wife and mother who sacrifices all for her chicks with whom she lives indoors in her little farmhouse—we find revealed in vivid detail the changed social position of the woman farmer in the United States between 1928 and 1957.

In response to subRosa's question, "Why are women like chickens, and chickens like women?" we might answer that an anthropomorphic reading of the *Little Red Hen* reveals that women share with chickens the status of being viewed as inessential, marginal, or even invisible parts of the farm economy. Yet this interpretation replicates the major shortcoming of twentieth-century feminism: its one-dimensional understanding of gender. With the powerful reorienting effect of intersectionality theory, feminists have come to see the category woman as reciprocally constituted with and by race, ethnicity, and nationality, among many significant identity categories (Crenshaw 2005). As women are involved in the chicken farming process, which extends beyond the farmyard to practices of chicken breeding, processing, and distribution, their experiences reflect their multiple overlapping identity positions, practices, and commitments.

From BEEF-EX to the Chicken Bone Express

In Ruth Ozeki's satirical novel about the global meat industry, *My Year of Meats*, an American documentary television producer's work on the Japanese television program "My American Wife!" propels her on a picaresque encounter with the ways that race, ethnicity, gender, and sexuality are cocreated through our international food system. Jane Tagaki-Little has been hired by the Beef Export and Trade Syndicate, a "national lobby organization that represented American meats of all kinds . . . as well as livestock producers, packers, purveyors, exporters, grain promoters, pharmaceutical companies, and agribusiness groups" to produce a series of television documentaries for the Japanese market (Ozeki 1998, 9–10). Each program would feature an American woman in her home environment extolling the virtues of American beef. Despite the fact that "the BEEF-EX people are very strict [and] don't want their meat to have a synergistic association with deformities. Like race. Or poverty. Or clubfeet," Jane's project morphs as she gradually discovers the environmental, social, and health costs of the meat industry for which she is working (57). Soon she finds herself subverting the program's beef-o-centric, blandly heterosexual narrative in favor of episodes featuring meats other than beef, seafood, and finally vegetarian meals served by a lesbian couple and a biracial couple with a family of multiracial adoptive children (57).

On the other side of the world, a young Japanese housewife named Akiko Ueno is discovering the horrific extent of her own sexual subjugation to her husband, the representative of the advertising agency in charge of the BEEF-EX promotion who is Jane Takagi-Little's "de facto boss" (41). The women's stories converge in the novel's final chapter, when Akiko shows up on Jane's doorstep in Manhattan's East Village, in an encounter filled with multiple, painful ironies. After being savagely beaten by her husband and discovering that she is pregnant, Akiko has left him to go to the United States so that the child inside her can grow up to be "an American wife" (318). In contrast, Jane is mourning a miscarriage, possibly an effect of her work on "My American Wife," since she was exposed during filming to the hormones used in cattle raising—to synchronize the estrus of a herd for easier artificial insemination" (261–262). When the two women meet, their interwoven stories testify to the varied ways that scientific interventions in animal breeding have also shaped the reproductive experiences of women.

The links between beef and chickens, and between the breeding and consumption of meat, that entangle Jane Takagi-Little and Akiko Uemo also intertwine gender and species in the management of reproduction. As Sarah Wilmot argues, such innovations in reproductive medicine were not merely the results of clinical research (Wilmot 2007). Arguing that "we should begin to question the privileging of the language and practices of genetics in popular historical accounts of modern reproduction in agriculture," Wilmot points out that for experts in reproduction, developmental embryology and reproductive physiology were as important as eugenics, while for laypeople the craft of breeding was of equal significance with the science of genetics (304–305). There is "ample evidence of the value of attending to the reproductive sciences that lie between craft and genetic science" (305). While Wilmot does not address poultry breeding, her argument has even more force there because the industrialization of animal agriculture was modeled on innovations in poultry farming, in particular the rationalization and standardization of chicken reproduction through the electric incubator and the electric hatcher/brooder but also through the move to mass confinement grow-out buildings. Moreover, the innovations extended beyond genetics and "hard" technologies to "soft" or social technologies as well, as William Boyd has pointed out: "While breeding and genetic improvement were clearly central vectors of technological change in making the industrial chicken, they were by no means the only ones. Intensive confinement, improved nutrition and feeding practices, and the widespread use of antibiotics and other drugs also represented important aspects of a larger technology platform aimed at subordinating avian biology to the dictates of industrial production" (Boyd 2001, 633). Such a subordination of biology to industrial production can also be found in the contemporary franchised fertility clinic, with predetermined identical office layouts, treatment protocols, and pharmacological regimes.

Adele Clarke has analyzed how innovations in animal agriculture have traveled to human reproductive medicine. She highlights in particular "two key interventions" that occurred in chicken farming and beef farming, respectively, producing major improvement in animal production (Clarke 1998, 159). The first intervention was the transformation of chicken farming from a female-centered activity carried out in the family backyard and barnyard to a "factory-based industry" carried out at a large-scale indoor location away from the family home. This was made possible by technical innovations such as electric light and kerosene heat for hen houses, which transformed the hen into "a mechanical oviduct"; electric breeders, which did away with the need for a hen's "maternal functions"; and artificial incubators, which made possible the production of both uniform and standardized chickens and eggs (159).[5]

The second important intervention, according to Clarke, was the introduction of artificial insemination to cattle breeding, which improved "the production of herd animals" (159). This made it possible to breed selectively to improve the herd, and the refinement of the whole range of practices required to carry out artificial insemination also improved the field of animal science in general. While Clarke focuses primarily on cattle breeding, it is important to point out that animal science itself was an outgrowth of poultry science, a thriving research field fully thirty years before the appearance of animal science as a research field in the 1940s. Since the early years of the twentieth century, the chicken had been seen as an ideal experimental model because of its small size, its relatively fast rate of reproduction and growth, and the crucial fact that its egg was accessible outside the body (Boyd 2001, 636). Indeed, women's reproductive timing has been managed and standardized toward the goal of increased fertility using techniques adapted from animal agriculture and based on work with laying hens.

A couple of landmarks can suggest the relay between chickens and women in reproductive terrains ranging from infertility treatment to gynecologic oncology. As early as October 5, 1937, the *New York Times* reported, "Science spanned 200 miles of land and water to mate a batch of barred rock hens with Rhode Island Red roosters by remote control, and then turned the experiment over to time and nature."[6] Characterized as "the first time chicken egg fertilization had been attempted between such widely separated fowl," this experiment entailed bringing sperm from Maryland in a special container that kept it at 40 degrees Farenheit, and then inoculating 300 hens with one-tenth of a cubic centimeter each (or, as poultry geneticist Joseph P. Quinn of the National Agricultural Research Center figured it, 15,000,000 spermatozoa).[7] The resulting eggs were then incubated at the Northeastern Poultry Producers Association exposition at Port Authority, New York City, in November 1937. Chickens were an important tool for researchers studying abnormal growth as well. Recall that as early as 1911, researcher Peyton Rous experimented with the viral introduction into healthy chickens of a sarcoma from an afflicted chicken. An effective vaccine was created for the chicken lymphoma known as Marek's disease in 1976 (Stone 3002, 34; and Steck and Haberstich 1976). Flash forward just thirty years and researchers were celebrating the chicken as a highly promising model for research on ovarian cancer (Giles, Olson, and Johnson 2006).

As Ozeki's *My Year of Meats* ends, Jane Takagi-Little is recovering from cervical cancer possibly caused by diethylstilbestrol, the hormone first used in agriculture to increase growth rates in chickens, while Akiko's reproductive life is just beginning. Akiko travels by train down to Louisiana for Thanksgiving, and then back up to Northampton, Massachusetts, where she finds an apartment of her own and a new American life, with the aid of the lesbian couple she visits who were Jane's first "spokeswives" for the meat industry (Ozeki 1998, 338). Juxtaposing African American families with a newly single Japanese woman on a train traveling north from Louisiana, the larger-than-life scene on the racially segregated "Chicken Bone special" dramatizes some of the more complex meanings chicken has held to African American women.

The legendary Chicken Bone special, also known as the Chicken Bone express, originated during times of African American migration out of the South to the North and West in search of employment, education, and equality. Routes became known as the Chicken Bone express because the travelers—forbidden by Jim Crow law or prohibited by penury from eating in the train luncheon cars—packed shoe-box lunches of fried chicken, throwing the bones out the train windows where they littered the ground, mute evidence of the African American migration. Akiko's guide for this journey on the Chicken Bone express is Maurice, the Amtrak attendant on the northbound train, who explains the lively scene in her railway car in terms that straddle the line between authenticity and stereotype: "Its called the Chicken Bone, Miss A-KEE-kow, because all these poor black folks here, they too poor to pay out good money for them frozen cardboard sandwiches that Amtrak serves up in what they call the *Lounge Car*, so these poor colored folk, they gotta make do with lugging long some home-cooked fried chicken instead, ain't that right now?" The passengers cheered. "Which one of you's got a piece of home-cooked fried chicken to share with Miss A-KEE-kow who's come all the way here from Japan? Give her a taste of some Southern hospitality now" (338–339). Maurice turns on the PA system and rallies the passengers into a chorus of "chicken bone, chicken bone, chicken bone" while he sings "She's a mighty old train / But she's runnin' just fine / An the folks who ride her, / They have a good time, / On the Chicken Bone, Chicken Bone, Chicken Bone special!" (339). Just as Akiko's sweet naiveté has provided the foil for Jane Takagi-Little's mordant realism throughout *My Year of Meats*, here too she accepts Maurice's sanitized message without realizing the deeper history of racial oppression it obscures: "*This is America!* she thought. She clapped her hands and then hugged herself with delight" (339).

To Psyche Williams-Forson, the tale of the chicken bone express speaks powerfully of the forgotten labor of African American women as waiter-carriers, engaged in "the trade of selling chicken, hot biscuits, coffee, and other foodstuffs to hungry train passengers who were eager to purchase their goods when trains stopped in their rural town" (Williams-Forson 2006, 1). Black women created a thriving informal economy selling chicken to travelers who were excluded by race or poverty from the amenities available to wealthier white travelers: meals, beverages, even beds in rooming houses for blacks. The tradition of waiter-carriers kept alive a specifically African American social identification, but more than that it provided black women with an economic, social, and even "metaphorical way of moving from one domicile to another" (115). For Akiko, still reeling from her own uprooting from her home in Japan, this waiter-carrier tradition is invisible. Her outsider status shields her from both the painful and healing truths beneath Maurice's smoothly sunny narrative.

"Chicken?"

While African American women played a culturally central role in the preparation and consumption of chicken, according to Williams-Forson, thus attaining
self-definition and agency despite the racist and hostile environment of the U.S.
South, they have also played a prominent and far less enabling role as mini-
mum wage workers in the poultry processing industry. Poultry processing
plants are "highly organized industrial structures for slaughter, disassembly,
and packaging of birds," and until the mid-1960s this highly dangerous, low-
paying job was filled mostly by black workers (Lipscomb et al. 2005, 1834). But
with the steep increase in the population of Hispanic immigrants (documented
and undocumented) in the South between 1960 and 1990, poultry processing
corporations started to employ more and more Hispanic workers. By 2005,
according to Steve Striffler, "about three-quarters of plant labor forces [were]
Latin-American, with Southeast Asians and Marshallese accounting for a larger
percentage of the remaining workers" (Striffler 2005, 112).

Latino/a workers were (are) frequently brought to the poultry plants illegally, recruited as undocumented immigrants by "the Hispanic Project," a secret
corporate "recruitment" scheme that advertised for poultry workers in Latin
American newspapers, attracting undocumented immigrants from Mexico,
Argentina, Cuba, and Guatemala (Fiege 2007). These workers, who in many
cases were brought into the United States by coyotes working under the table
for the major poultry corporations, were bused into Mississippi from Florida
and Texas, where they were hired as line workers by the poultry plants. Their
situation as new employees was dire: speaking no English, they often had no
legal status and no awareness of their rights as workers (Fiege 2007). John
Fiege's moving documentary, *Mississippi Chicken*, documents the lives of poultry workers in Canton, Mississippi, as they deal with brutal employment conditions and a hostile local environment.

Nominated by the Independent Filmmaker Project (IFP) for a Gotham
Award for "Best Film Not Playing at a Theater New You," *Mississippi Chicken*
paints a complex picture of the interaction between African American and
Latino/a communities as they negotiate a rapidly changing employment picture
for low-wage workers. There are quietly harrowing interviews with Latina
women who have been injured in poultry processing, particularly one quiet
young woman whose finger was trapped, partially amputated, and the remaining digit flayed to the bone when it caught in the shrink-wrap line. But the
film focuses on the tale of Guillermina, from Mexico, and her newly arrived
fourteen-year-old daughter Rosario, or Charro. Narrated by Anita Grabowski, a
worker's rights advocate who meets the two women when she is starting to
organize a center for poultry workers in Mississippi, the film follows the painful
arc of Charro's story. We learn that for these Latina women, a job in the poul-

try industry, which once held out the hope of stable work and a viable life for themselves and their families in the United States, has come to mean the high risk of injury and the near-certainty of low-wage exploitative labor.

The film's joyful opening scene shows preparations for Charro's fourteenth birthday celebration, as her family works alongside an African American farmer to slaughter a pig for the birthday meal. A later scene captures Charro's hopeful early days of high school; although she declines to predict her future, she firmly assures the interviewer she will never work in the poultry industry. The film's initially celebratory mood fades quickly, however. On her next visit to Guillermina, Anita (the worker's rights organizer) learns that Charro has been raped by the twenty-year-old man she has secretly been dating. When Guillermina found this out, she took her daughter out of high school and shipped her off to live—at less risk, she hopes—with relatives in North Carolina. On the film's final visit to Guillermina, several months later, Anita learns that Rosario (Charro) has left school for good and is working in a poultry plant.

The conclusion to Rosario's sad story happens off camera, but with visual and verbal economy Fiege's film makes very clear the charged and complex relationship between these particular Latina women and the chicken industry of Canton, Mississippi. One brief scene speaks volumes. David, the Catholic brother who has been helping to organize the poultry workers' center, arrives for dinner, bringing food with him. "Chicken?" Guillermina sniffs, with evident disapproval. "I don't buy from Tyson," she says. "Tyson steals" (Fiege 2007).

One Hen

Who has access to the fruits of one's labor is a central theme of the children's story with which I began, *The Little Red Hen*, whose anthropomorphic protagonist epitomizes the close connection between women and chickens in our agricultural imaginary. Another children's book published to some critical acclaim in 2008 also concerns the same theme, yet this story about a small reddish brown hen also demonstrates the other issues that must complicate any exploration of the role of women in agriculture. Katie Smith Milway's *One Hen: How One Small Loan Made a Big Difference* might be called documentary fiction, because it is based on the true story of Kwabena Darko, "a real boy from Ghana's Ashanti region who really did lose his father and have to help his mother support his family" and it relates how the purchase of one small brown hen led to a successful venture into microcredit lending (Milway 2008, 28). Unlike both versions of *The Little Red Hen*, where the advocacy of a cooperative sharing of work and profit seemed to extend no further than the boundaries of the farmyard, *One Hen* is explicit about the importance of a broadly framed notion of community, both to the characters in the story and to its readers.

Young Kojo, the only child of a single mother in a small village in Ghana, has had to quit school upon the death of his father to help his mother make enough money selling wood at the market so that they can support themselves. When Kojo is able to borrow a few coins from the fund of money shared by the twenty families in their village, he gets a good idea. He travels to the neighboring village and buys a little red hen, which he carries home in a basket on his head, and settles it beside his bed-mat in an old laundry detergent box. To his delight, the hen not only lays eggs but also survives on the fruits and loose grains that Koto scavenges from the market. Kojo's "good idea" provides more than sustenance for his mother and himself: it introduces him to the life of an entrepreneur. Within six months Kojo has three hens that lay enough eggs for both of them to eat an egg every day and sell the rest at the village market. Within a year, he has saved enough money that he has options: "Maybe he will use his egg money to build a fine wooden chicken coop. Maybe he will buy some things his mother needs, such as a new water bucket and a good knife. Or maybe he can pay for something he's been dreaming of: fees and a uniform so that he can go back to school. 'Your eggs have made us stronger, Kojo,' says his mother. 'No go to school and learn . . . for both of us'" (11–12).

Kojo takes the long-term option, "for both of them." He goes to school and wins a scholarship "to an agricultural college to learn more about farming. His mother will care for his chickens while he is away" (15). By the time Kojo reaches adulthood, he feels ready to take the biggest risk of his life: "He will use all the money he and his mother have saved to start a real poultry farm" (16). He goes to the capital city of Accra, to the bank's headquarters, and finally meets with the busy bank president. As Kojo tells his story, the dubious bank president smiles at the story of little Kojo's business acumen: "This is not a story he hears every day. He smiles and nods—Kojo will get his loan" (16).

The tale takes on the familiar cast of any successful business narrative. With a thriving business, Kojo is able to marry, a teacher who shares his dream of [male] self-betterment: she "has many stories about boys just like Kojo once was—boys who want to learn and who have big dreams" (19). Kojo's poultry business thrives so much that he has to hire other workers to perform tasks that seem assigned by gender: "Men come to feed the chickens and clean the coops. Women collect the eggs and pack them in boxes." Soon "Kojo's farm is . . . the largest in Ghana. And his town has grown, too. Some people come to find jobs on the farm and build homes for their families. Others come to the town to open shops and sell wares to the workers" (23). Kojo even lends money to Adika, a young woman whose family has worked for years on his poultry farm. In a plan that recalls *The Little Red Hen*, Adika hopes to combine her savings with the loan to "buy a mechanical grain mill and start a small business helping families turn their grain into flour" (23). By the story's end, "Kojo's poultry farm [has become] the largest in all of West Africa" (24). A ripple-effect from the taxes Kojo pays—

along with "his workers and the shopkeepers who sell his eggs"—has led to improvements in "roads, schools, and health clinics across the country," and even to "the port at Accra where ships from many countries come to trade" (24). The children's book ends with a vision of a secure nation, trading globally, with a vibrant infrastructure and a secured patriarchal lineage, all produced by one small loan to buy one red hen: "One more egg truck drives away, and Kojo looks down at his youngest grandson. The next time the boy asks Kojo where an egg will go, Kojo will say, 'To your future, my child'" (24).

Such an expansive perspective on community is important not only to the people in the tale of *One Hen* but also to its readership, who are asked to consider the good of their global community. "What can you do to help?" the author asks in an afterword, and then responds to her own question, "There are a number of organizations in North America that donate money to 'village banks,' such as Sinapi Aba Trust in Ghana [founded by the Kwabena Darko, the model for *One Hen*'s Kojo], BancoSol in Bolivia, and Grameen Bank in Bangladesh. These and similar groups in other countries lend people money to start or build a small business and pull themselves out of poverty" (30).

The microcredit loan plan for sharing investment costs and thus expanding social benefits seems like a far more appealing solution than the chorus of "Not I!" that greeted Little Red Hen's appeals, or even than her deliberate decision not to share the bread but to eat it alone or to feed it all to her chicks. The Sinapi Aba (Mustard Seed) trust that the "real Kojo," Kwabena Darko, started after completing his poultry science studies at agricultural college in Israel makes small loans to groups of people, enabling them to start their own businesses. Moreover, unlike the patrilineal model that seems built into Kojo's story, "about 90 percent of the women who receive loans are women like Kojo's mother" (29). Yet it seems curious that the children's story focuses on the more unusual 10 percent of male loan recipients, devoting only passing notice to the effect of the small loan on the women of the village. We learn that Kojo's mother cares for his chickens while he is away at agricultural college and then comes to live with him and his wife in their new cinderblock house, freed forever from the need to sell wood for a livelihood. Only Adika, of all the women in the story, shares Kojo's entrepreneurial vision and path. If we reexamine the story in light of a contemporary critique of microcredit lending, we will see that the strategy of a microloan to support small businesses may have a far more complex impact if the loan recipient is female.

Indeed, such microcredit loan programs may not be the definitive answer to the problem of global poverty. In a fine-grained analysis of the emergence of microcredit lending in Nepal, where the overwhelming majority of such loans come to women, Katharine N. Rankin concludes that only if the lending programs enable the collective identification of the female loan recipients "as women, across polarizing social differences" are they actually empowering for poor women (Rankin 2001, 31). Rankin concedes that microcredit has "politically progressive potential . . . to mobilize local resources for social change," and she disagrees with the most critical interpretation which would view microcredit as a pernicious governmental strategy "in its appropriation of feminist languages of empowerment and solidarity to alternative (and fundamentally conservative) ends" (31). However she draws attention to the contradictions within women that shape the potential effectiveness of microcredit loans: "Caste, ethnicity, and class are obvious examples of the kinds of social distinctions that might structurally preclude women in some social locations from viewing their interests in solidarity with women in others" (31). By focusing on the story of Kojo, *One Hen* avoids these complications, instead limiting the notion of community solidarity to the familiar and seemingly natural patriarchal family within which this young boy takes care of his mother, his wife, and ultimately his male descendents.

Appreciation of the history of the globalization of poultry farming will also illuminate the complexities swept aside in the simple optimism of Kojo's story. Microcredit loans introduce more than welcome capital to a poor community, Rankin argues; they also import into the global south the neoliberal business models of the West. Kojo's evolving poultry farm is an instance of this because it reflects the gendered economies of scale that Western industrial poultry farming introduced to agriculture in the United States as chicken farming moved from a female barnyard economic supplement to an agricultural industry in its own right.

Kojo has learned well the poultry science lessons of his agricultural college. While his initial schooling taught "practical lessons for country life . . . how to use chicken manure and compost made from garbage to fertilize soil and grow vegetables," as the farm grows to be the largest poultry farm "in all of west Africa," its inputs and outputs are increasingly segmented. Clearly those chickens no longer feed on barnyard scraps, and the sheer volume of manure they would produce suggests that they no longer fertilize nearby vegetables. Moreover, the "thousands of eggs a day" that his chickens lay circulate beyond the local community, feeding families in such adjoining countries as Mali and Burkina Faso. A caption on an illustration celebrates the effects of the farm's productivity: "This is the country that grows as businesses like Kojo's and Adika's prosper" (25).

The development of Kojo's poultry business, guided by his work in poultry science at the agricultural college, not surprisingly echoes the gendered changes inherent in the move away from the family poultry management characteristic of so-called developing countries to a Western-style intensive poultry business. "Women are the main poultry owners in developing countries, though there are variations within and between countries," points out E. F. Guèye, adding, "on the whole, women's involvement in poultry farming tends to decrease with increased level of intensification" (Guèye 2005, 41). Although Kojo abandons the message of sustainable growth that he learned in his local village school along with the family poultry farm model it incorporated, the childhood vision of entrepreneurial expansion that won him the loan in the first place is retained and compounded.

Why Are Chickens Like Women and Women Like Chickens?

In a pamphlet that combines adbusting play with poultry advertising, fact-lines and narratives of the linked projects of eugenics and animal agriculture, parodic rephrasing of documents from the pharmacology and poultry industries, and a statement from the DelMarva Poultry Justice Alliance speaking for the rights of poultry workers everywhere, subRosa Art Collective has suggested that women are like chickens, and vice versa, because they share the position as sex workers for global industry: they both produce eggs, are both subject to bodily regulation and regimentation, and are both being shaped biomedically and socially to be the object of male sexual and economic desire. While there is much to agree with in subRosa's pamphlet, it neglects the most basic context in which women and chickens have historically coexisted in mirrored relation: the institution of agriculture. By examining several works of art that portray the relation between women and chickens from a perspective attuned to the intersecting influences of race, ethnicity, and nationality, we have arrived at a more complex and situated portrait of woman's role in one realm of agriculture: the practice of poultry farming. And perhaps we have a better understanding of the liminal nature of that long-beloved character, the little red hen.

Hybridity

"Culture and history must be rethought with an understanding of their inextricable, if highly variable, relation to biology."

—Lennard J. Davis and David B. Morris, "Biocultures Manifesto"

"[One] cannot fully understand cultural practices unless 'culture,' in the restricted, normative sense of ordinary usage, is brought back into 'culture' in the anthropological sense, and the elaborate taste for the most refined objects is reconnected with the elementary taste for the flavors of food."

—Pierre Bourdieu, Distinction

In February 2008 I spent several days in intense conversation with artist Koen Vanmechelen—at his kitchen table in Meeuwen-Gruitrode, Belgium, where we sat surrounded by cabinets on which perched taxidermied chickens; on the bright red couch in his living room, under the vivid photograph of a white rooster with a bright red comb; at the dining table, where from a sideboard stared three stuffed lizard-chicken hybrids; and finally in his Toyota SUV as we drove at high speed through Belgium and the Netherlands. It was a whirlwind visit; Vanmechelen had just returned from presenting his work at the World Economic Forum in Davos, Switzerland, and during my visit he was negotiating with a collector in Belgium to produce an installation for his house and grounds; overseeing his first solo show, "The Chicken's Appeal," which had just opened at the Museum Het Valkhof in Nijmegen, the Netherlands; meeting with prospective collectors at his studio in Hasselt, Belgium; and preparing for shows in Bern and Berlin later that spring. Nevertheless, he graciously welcomed me to the house he shares with his wife, a nurse and hospice administrator, and he gave me a tour of the grounds, which include a well-populated poultry yard and outbuildings housing ranks of breeding cages, fenced in fields where alpacas graze, several caged kookaburras, a resident dog, and several cats.

I had been trying to track down Vanmechelen for several years, ever since I heard about his Cosmopolitan Chicken Project, a multiyear, multimedia art-

FIG. 21. The Cosmopolitan Chicken Project.
Used by permission of Koen Vanmechelen.

work with the goal of creating a "cosmopolitan chicken," or what Vanmechelen called a "hybrid super bastard." With the help of e-mail friends in Australia and the United Kingdom, I was finally able to track down Koen and arrange an interview. He met me at the train in Brussels, and as we struggled to converse in the noisy Greek taverna we had chosen for our dinner, I was introduced to the vision motivating his art. As I learned that night, and during the several days that followed, Koen Vanmechelen reclaims two nineteenth-century obsessions—chicken breeding and the concept of hybridity—to make a twenty-first-century intervention into our understanding of the complex meaning of culture. His Cosmopolitan Chicken Project offers a broad interpretation of chicken culture, comprising a wide range of issues: taste, breeding, origin, language, domestication, nation, morphology, normativity, and species.

An energetic and charismatic presence with his dark eyes and long dark mane of hair, Vanmechelen was born in 1965 in the Flemish village of Sint-Truiden, to a fashion designer mother and a sculptor father who was a teacher of classics (Simons and Kierse 2003). Taste, in the dual senses that Pierre Bourdieu enlists in his work—normative and anthropological, sensory and discriminating—seems to have been a passion for Vanmechelen from his earliest childhood. He was fascinated with chickens, keeping them behind the house when he was young and building wooden bird houses, cages, and later wooden sculptures of chickens. Rather than joining most of the students at his school in the mathematics curriculum, he followed the Greek and Latin course at his public high school, studied hotel and restaurant management for one year, and then became a chef and pastry cook, working his way up to a two-star restaurant, La Feuille d'Or. Perhaps following in his father's footsteps, he took particular interest in sculpting chocolate. Yet as he aspired to create a new dessert plate every day, he showed more signs of being a talented cook than of an artist-in-the making.

But by the late 1990s his interests were shifting, for Vanmechelen decided that he would attempt to improve on the two most highly sought-after chickens served in his region: the Belgian Mechelse Koekoek and the French Poulet de Bresse. Although both were thought to have delicious flavor, to his way of thinking they were not still fully "finished." With taste foremost in his mind, he decided to hybridize them to improve their taste, since the Melchese Koekoek seemed too fatty while the Poulet de Bresse was too tough. Vanmechelen had to go through some complex negotiations to acquire specimens of the heavily guarded French Poulet de Bresse, which legal regulations and political tensions have maintained as a separate and distinct breed. "He traveled to a heavily guarded Bourg-en-Bresse, where he managed to wheedle a few specimens of the French national pride out of their owners, crossing his heart that he had no intention of using them for illegal breeding" (Simons and Kierse 2003, 25). A Mechelse Koekoek was far easier for this Flemish cook to obtain, and having done so he set up a breeding operation in the garden of La Feuille d'Or, the two-star restaurant that was his current employer. He even created a hatchery in the restaurant's basement, where among the dozens of chicks that hatched from this new breeding project appeared one bluish chick—the first representative of a new breed Vanmechelen decided to name the Mechelse Bresse. Yet this breed would not be a new culinary resource because the legal regulations around copyrights and patents that he encountered for both closely guarded breeds made a continuing breeding program prohibitive. The search for savor gave way to its aesthetic double: "taste" (7–11).

Why did a simple exercise in breed improvement face such obstacles? It was a simple matter of national pride. The two chickens Vanmechelen chose for his initial cross possessed special legal standing as a breed because the nations in which they were grown claimed them as national icons. In 1957 the Assemblée Nationale in France passed a law that gave the title *appellation d'origine contrôlée* to the Poulet de Bresse, while in 1958 the prized Flemish chicken, the Mechelse Koekoek, had its debut presentation at the Brussels World Fair, following the long tradition of such fancy breed exhibitions. More than culinary taste was at stake in Vanmechelen's challenge to the property rights of growers and exhibitors by crossing these two iconic chickens: nationality and linguistic identity were also in play.

As a Dutch-speaking Belgian at home both in his childhood Flanders and in the broader Western tradition for which his classical education prepared him, Vanmechelen well understood how the collision of historical, national, and linguistic forces could tangle and trouble any notion of biologically based identity.[1] Conflicts over linguistic and cultural identity are fundamental to the Cosmopolitan Chicken Project. With the decision to cross the famous Flemish Mechelse Koekoek and the French Poulet de Bresse, Vanmechelen challenged the tradition of regional and linguistic loyalty dating back to clashes between France and French-speaking Wallonia in the south and Dutch-speaking Flanders in the north catalyzed by Belgian independence in 1830. Historical disparities, both linguistic and economic, have traditionally separated the southern region of Wallonia—home of the poorer, French-speaking "Walloons"—from the wealthier, Dutch-speaking region of Flanders to the north, where Vanmechelen lives and works. In 2007, after an electoral majority ratified the plan of the Christian Democrats and Liberals to create a Belgian coalition government, "both sides then split into Dutch and French-speaking parties with the Dutch-speaking Christian Democrats and Liberals demanding more autonomy for the northern Flanders region" (Amies 2007). The economic and social disparity between these linguistically divided regions has led to an impasse: "The problem is that we are still stuck in an electoral situation where any movement by one side or towards an agreement is seen as a sign of weakness. . . . It's like a game of political chicken. Who will blink first?"[2]

In a sense, Vanmechelen's art can be understood as a contribution to the very same battle of wills between the French-speaking Belgians and the Dutch-speaking Flemish, and yet he attempts to get the other to "blink first" not through political aggression but through visually arresting art. Beginning with his creation of the Mechelse Bresse, Vanmechelen next created a one-man show, "Installatie AI" at the Cultural Center in Hasselt in 1997, where he charted the breed crossings he accomplished in stunning photographs of paired chickens. From that first exhibition on, Vanmechelen's project developed over the course of more than eleven years to include representations of chickens in a wide range of media—watercolors, tempera, graphite, glass, gold, egg yolk, feathers, skin, and living organisms—and a remarkable variety of methodologies—painting, drawing, sculpting, glass-blowing, taxidermy, video, CAD-CAM (computer aided design and manufacturing), performance art, and conceptual art. By the spring of 2008 when we met, he had created installations related to his project of chicken hybridizing in sites across Europe, India, and Africa with displays that included the breeding pairs housed in long chicken runs labeled with a color photograph of the chicken indicating its name and sex as well as paintings, sculptures, blown glass, and video representations of chickens.

Extensive in time and space, this work of art disrupts conventional understandings of what chicken culture might mean, whether in terms of morphology, aesthetics, or politics. Unlike the scientifically based portraits of superior and inferior breeds that guided poultry breeders such as Jean Pagliuso's father, Vanmechelen's chicken representations do not aim at creating a perfect purebred chicken. And unlike the aestheticized chicken portraits in Jean Pagliuso's "Poultry Suite," they do not aim to give us culture in the social sense either, with overtones of female high fashion and idealized female beauty. Instead, Vanmechelen's work resists a standard of excellence, whether established by the classic poultry breeder's volume, *Standard of Excellence in Exhibition Poultry*, or by the studios of elite fashion designers. Deliberately breeding hybrid chickens, he also challenges any notion of a norm, for his installations include the malformed chickens that occasionally also hatch from time to time, a result that he sees as an equally important contribution to the project of creating the "hybrid super bastard" (Robinson 1912a). An installation titled "Genetic Genius," part of the "Secret Garden" exhibition at a former psychiatric hospital, thus included a taxidermied assemblage chicken with two heads, formed from the bodies of a chicken that had been born with a hump and another with a lame leg.[3]

Vanmechelen's art also challenges the fixed boundary between nature and culture. Unlike the naturalist portraits of different chicken breeds representing chickens in field or barnyard, such as Robinson's classic poultry engravings and the illustrations in Tegetmeier's *Poultry Book*, Vanmechelen's chicken photographs are grouped with works in other media—from wood to glass, taxidermy, and gold. It is as if the form of the species *Gallus gallus*, like the form of an individual, were polyvalent, pliable, open to negotiation, decoration, and reimagination, beyond the simply instrumental impulse of improvement. The chicken's discrete identity is cross-cut, fissured, or combined with another bird, another medium, or even another species, to transgress not only art space but also science time. In a series of visual and tactile images, Vanmechelen also seems to be reversing the evolutionary development from reptiles to birds. This temporal spooling was strikingly apparent in the photographs, watercolors, and pencil drawings of chickens with the heads of lizards, and the taxidermied chicken-lizard and chicken-snake assemblages—hybrids that stared at me from the dining room sideboard as we sat talking in the kitchen (Vanmechelen 2007; and Schoor and Coucke 2008, 50).

During my time with Vanmechelen, I saw how his Cosmopolitan Chicken Project (CCP) articulates his own powerfully personal aesthetic response to interdisciplinary questions of origin, taste, breeding, domestication, nation, normativity, and species. I came to admire this response for its catholicity, iconoclasm, and intellectual daring. Yet after the trip to Belgium I remained undecided about what kind of an intervention this project made into the concepts that form the core of the project: cosmopolitanism and hybridity. The question asked by a visitor to one of Vanmechelen's exhibitions suggests that we can tease out some of the implications of the CCP if we assess (and then disregard) its relation to scientific knowledge. The visitor described the exhibit's "pseudoscientific breeding and hatching area, in which viewers can watch the pattern of development from egg to chicken" and went on to conclude that the exhibit offers "a metaphor for art" (Simons and Keirse 2003, 28). A similar default to art seems to be suggested by the assertion, presumably made with Vanmechelen's permission, that appears on the Web page for one of the two galleries that represent his work, Venice Projects, Ltd.: "All the various artistic forms that he uses are connected by two subjects central to his work: the chicken and the egg, chosen as metaphors of man and of life."[4] Yet this retreat from science to the realm of art does not satisfy. How can an actual breeding and hatching area— where the process of chicken reproduction runs its course and an incubator and hatchery enable the resulting eggs to become chicks—be seen as merely "pseudo-scientific"? Such a label assumes that only training in a science discipline could authorize Vanmechelen to create art that not only engages with but actually participates in science. Claiming the CCP as art rather than science demonstrates that in this case at least the "two cultures" division is the result of artistic well as scientific boundary policing (Snow 1993).

Cartography and Hybridity

A close look at "*Gallus gallus bankiva* (2000)," one of Vanmechelen's most conceptually direct art works, offering a summa of his vision, suggests that it would be wrong to write off his use of science as merely metaphoric. The initial image in the catalog for Vanmechelen's 2008 solo show at the Museum Het Valkhof in Nijmegen, the Netherlands, this work makes direct reference to cartography

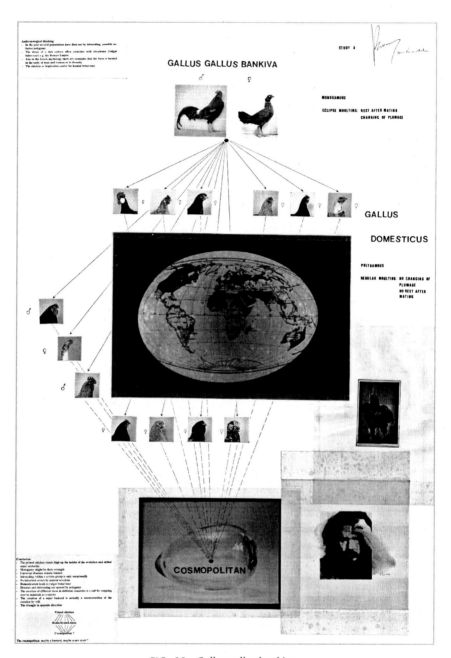

FIG. 22. *Gallus gallus bankiva.*
Used by permission of Koen Vanmechelen.

and evolutionary biology, two scientific fields intimately involved with the concepts of hybridity and cosmopolitanism so central to the vision of this twenty-first-century artist-cum-chicken breeder (Schoor and Coucke 2008, 50). At the top of the image, the phrase "*Gallus gallus bankiva*" provides the scientific label for the photographs of a breeding pair of Indian jungle fowls positioned beneath it. Lower still, what appears to be an image of the globe superimposed upon a photograph of a brown egg occupies the central position.[5] This allusive geographic/biological construction is surrounded by a series of photographs of different chicken breeds, each labeled male or female. Breed portraits, which resemble theatrical head shots, are each linked by one of a series of radiating lines to the male progenitor above (*Gallus gallus bankiva*) and to a glass egg below, labeled "Cosmopolitan." Just off to the right is an image of a stocky black chicken, resembling the Lisson Gallery (London) image of "*Gallus gallus bankiva*" (2000) and beneath it the close-up of an intent Koen Vanmechelen, dark hair and beard, hand to his forehead as if thinking deeply (Simons and Keirse 2003, 21). Off to the lower left we see an enigmatic list labeled "Conclusion" including a diamond-shaped graph where diamond lines link "primal chicken" at the top to "Cosmopolitan?" at the bottom, with "Domesticated races" neatly caged in the middle. Beneath the diamond graph is a question: "The cosmopolitan may be a bastard, maybe a new start?"[6] And to the top left we find a boxed list setting forth four principles of "Anthropological thinking." Points one through three have a scholarly resonance: the role of inbreeding in population die-off, the relationship between cultural decline and decadence, and the resonances in Greek mythology of unity or disunity between men and women. But point four speaks directly of issues that preoccupy the artist, as they did Darwin: "The chicken as inspiration source for human behavior."[7]

To move beyond seeing Vanmechelen's chickens as simply metaphors for art and instead to discover what he is trying to express about chickens, human beings, and life itself, we must look more closely at Vanmechelen's encounter with the science of cartography. With close examination, the image superimposed on the egg at the center of "*Gallus gallus bankiva*" resembles a specific style of map: an equal-area, pseudo-cylindrical Mollewide projection, developed in 1805 as an advance upon the reigning Mercator projection (Furuti 2009). This style of mapping the earth was invented in 1569 by Flemish geographer and cartographer Gheert Cremer (Latinized as Gerardus Mercator), also known for his map of Flanders (1540) and the celebrated map of India, the "India Orientalis," that he created with Jocundus Hondius for their 1606 *Atlas sive cosmographicae meditations* (Varadarajan 2005). Mercator's creation, a rectangular technique of mapping the globe, represented a major advance in nautical navigation. However, it had one major drawback: it distorted the size and scale of the continents, making Europe seem far larger than it actually is. The story of the politically and socially charged debate over the Mercator projection, which became the focus of a controversy about the ideological and political nature of cartography that has lasted to this day, can illuminate Vanmechelen's intentions in this cartographic decision. Critiques of the Mercator projection frequently turned on the topic of the "political complicity of cartographers in the creation and maintenance of a Eurocentric world-view" in which "the Mercator projection was held up as a symbol of its age, and as the inevitable icon of global Europeanization. Its area distortion, in favour of higher latitudes, and its European centring were crucial to this argument" (Vujakovic 2003, 63).[8]

The decisive challenge to this problem of cartographic bias came with the introduction of the Peters projection in 1983. The Peters projection was designed to address two points, according to Cambridge University geographer Peter Vujakovic: to argue "first . . . for the complicity of mainstream cartography in European imperialism . . . , and secondly, [to promote] his own, supposedly egalitarian and objective 'new cartography'" (Vujakovic 2003, 62). Yet in trying to provide an "alternative global concept, by promoting his own projection," Peters was participating in a much broader debate about the ideological power of cartography, pitting those who argued that maps can serve the interests of power elites against those who argued that mapping can be strategically used by the disempowered to subvert dominant groups and ideologies (Vujakovic 2003). Ironically, Arno Peters's crusade against the claim to cartographic objectivity fell victim to the same problem, so that the Peters projection held a paradoxical position in the disciplinary debate: "While [Peters] accepted and publicized the value laden nature of traditional cartography, he attempted portrayed [sic] his own cartographic project as scientific and objective" (Vujakovic 2003, 62). By 1989 the issue had become so contentious that seven North American geographic organizations issued a resolution calling for a ban on rectangular coordinate maps.[9]

Vanmechelen invokes the full range of contemporary conversations about the ideologically laden nature of maps by choosing an illustration of the Mollweide projection, a compromise solution to the fierce debate between the Mercator and Peters projections, for his foundational image of "*Gallus gallus bankiva.*" He positions himself between those who critique them for enforcing the "subliminal geography" and those who find in the process of remapping a resistant strategy with which the global South can counter northern dominance.[10] As Indian journalist Siddharth Varadarajan has put it, "inherent in the cartographic imagination—in the very act of rendering intelligible the world with lines and shapes on stone, parchment or vellum—is always and everywhere an attempt to fashion new social boundaries and domains from the arid reality of geography."[11] Koen Vanmechelen's image, with its central globe and ample white space, provides plenty of room for playing with new social boundaries and domains. Yet in choosing a cartographic position in between the two disputed maps, it also invokes the very same question Varadarajan raises. Could

an unbiased map—a map that is completely the product of scientific objectivity—ever be created? Would it be possible to represent the world with no loyalty
to any specific place or state, as if one were a citizen of the world?

At issue here is the possibility of cosmopolitanism, a cultural and social
concept derived from the Greek word *kosmopolitês* ("citizen of the world").[12] To
social theorists and humanists, the term *cosmopolitanism* has been the subject
of some contention. For Immanuel Kant, cosmopolitanism indicated a community comprised of all humans, united by mutual respect, a common set of moral
principles, and the international relations of commerce. As he outlines it in *Perpetual Peace: A Philosophical Essay* (1795), "The rights of men, as citizens of the
world, shall be limited to the conditions of universal hospitality. It is not a right
to be treated as a guest to which the stranger can lay claim—a special friendly
compact on his behalf would be required to make him for a given time an actual
inmate—but he has a right of visitation. This right to present themselves to
society belongs to all mankind in virtue of our common right of possession on
the surface of the earth on which, as it is a globe, we cannot be infinitely scattered, and must in the end reconcile ourselves to existence side by side" (Kant
1917, 137–138).[13]

As modernist thinkers expanded the term "cosmopolitan" to include qualities of international scope, sophistication, and social and economic privilege,
they began to mark a certain implicit elitism, for example, in understanding
that cosmopolitanism involves a "right" of presenting oneself to society.
Reframing the concept as the new or actually existing cosmopolitanism, such
scholars as Gayatri Spivak and Kwame Anthony Appiah have argued that the
new cosmopolitanism incorporates "the principle of the modern state writ
large," and thus ignores the economic impact of globalization, whose structural
arrangements privilege the wealthy (and thus cosmopolitan) and punish the
poor (and thus provincial) (Fine 2003, 462).[14] This new critique views cosmopolitanism as a strategy of economic and social domination, a sort of societal
equivalent to the process of political and economic standardization necessary
for the dissemination of international corporate capitalism from the global
North throughout the global South. As Robert Fine describes this new perspective on cosmopolitanism, "In this conception, the unity of the cosmopolis, no
less than the state, cannot be conceived apart from the will of individuals who
retain their particular interests, identities and rights, for its principle is to
harmonize species-wide universality with the freedom and well-being of individuals" (462). Fine dissents from this view, arguing that since "the new cosmopolitanism encounters the same kind of antinomies that it exposes in its
critique of modernist politics," as a creation of Western modernity it is subject
to the same limitations (464–465).

More recently still, sociologists Ulrich Beck and Natan Sznaider have
returned to the term cosmopolitan to redeem it of its high-modernist overtones

of European elitism, internationalism, and even economic globalization by recasting it as an "empirical investigation of border crossings and other transnational phenomena" (Beck and Sznaider 2006, 1). They argue that rather than thinking of cosmopolitanism as a style or attribute—traits we might think of as being particular to a Paris fashion model rather than one of August Sander's village women, for example—we should think of it as a method of inquiry, a productive way to investigate identity and community in this postmodern era. Rather than perpetuating dualities, they argue, cosmopolitanism as a methodology does away with all familiar binaries—global and local; international and national; us and them. It gives rise instead to new institutions and formations requiring "conceptual and empirical analysis" (3). Reconceived as a transdisciplinary research methodology and agenda, their cosmopolitanism incorporates the major social science disciplines, from geography and anthropology to sociology and social theory.[15] As a method of inquiry, they argue, cosmopolitanism requires us to reexamine "the fundamental concepts of 'modern society.' *Household, family, class, social inequality, democracy, power, state, commerce, public, community, justice, law, history, memory and politics*" (6).

What kind of an intervention does the Cosmopolitan Chicken Project make in the notion of cosmopolitanism? Arguably, its aim is to reframe the notion itself, via practice and representation. The practice at the core of Vanmechelen's project, the deliberate creation of so-called cosmopolitan chicken through the technique of crossing pure, iconic national chicken breeds to create an artist-produced mongrel requires viewers to reexamine many of those "concepts fundamental to modern society" itemized by Beck and Sznaider. Seen in this light, the generations of cosmopolitan chickens, bred in his backyard as part of his household, emblematically offer a challenge to the system of class and linguistic inequality between Flemish and French speaking Walloons. From creation through exhibition as part of his art installations, these chickens required him to negotiate state laws and regulations governing commerce, the exercise of power, the meaning of community, and the protection of safety. A typical instance of such negotiations occurred in 2007 when Vanmechelen attempted to show his birds at the Basel Art Show in Miami Beach Florida. Due to regulations on the importation of live poultry enforced due to fear of the global spread of high pathogenic avian flu, or H5N1, he was prevented from exhibiting the chickens themselves. Instead, he was only permitted to exhibit their representations: photographs of his Cosmopolitan Chicken Project.

We might understand the CCP as a performative inquiry into relations between landscape, language, and species, with the goals of memorializing the Flemish community, extending its claim to cultural and historical memory, reinforcing its political solidarity, and thus rendering justice. And yet in the face of the prominent role played by biology and nature in his art, we have to do more than understand the chicken breeding project as "a metaphor for art" (Simons and Keirse 2003, 28). Precisely in its combination of culture and nature, sociology and biology, the CCP participates in what Vicky Kirby has argued is one of the most far-reaching but underanalyzed results of the association of cosmopolitanism with cultural change and mobility: "the cementing of the sense of an outside of culture that is wanting in these very qualities" (Kirby 1999, 21). As Nigel Clark paraphrases her position, "The cultural turn of the humanities and social sciences has tacitly bolstered western thought's time-worn binary of active, articulate culture and silent, docile nature" (Clark 2002, 107–108).

If we recall that Kantian cosmopolitanism consisted in shared morality and commerce grounded in our common right of possession on the finite globe, we can begin to delineate the contradiction between the human cultural intervention that is the goal of the Cosmopolitan Chicken Project and the biological, natural practices of cross-breeding by which it aims to accomplish its goal. The pivot point, of course, is hybridity—Vanmechelen's aspiration to create the "hybrid super bastard." Hybridity theory has been advanced by critics of Kant's universal cosmopolitanism such as Homi Bhabha and James Clifford as a critique of the imperialist ambitions and colonializing practices of the universal notion of culture on which it was founded (Cheah 2006, 80–119). As Bhabha has observed, hybridity "enables a form of subversion . . . that turns the discursive conditions of dominance into one of intervention" (Bhabha 1994, 112). Not only does cultural hybridity fuel the contemporary subversion of existing political and social hierarchies but it might also have driven the transition from the Mercator projection, criticized for reinforcing Eurocentric oppression through the relative distortion of its land masses, to its alternatives, changes with culturally significant implications. One such alternative, the extreme Peters projection, was introduced to English readers in the New Internationalist in 1983 as part of a two-part special focus on "political geography—on how people use and experience the places in which they live. Part 1 . . . included reports on communities as diverse as Australian Aborigines, African school teachers and European Gypsies."[16] Another alternative is the compromise Vanmechelen chooses to center his "Gallus gallus bankiva": the equal-area Molleweide projection of the globe.[17]

Whether radically or gently, Vanmechelen metaphorically inverts cultural geography when he crossbreeds the iconic French chicken with the iconic Flemish one, overturning French cultural, economic, and linguistic hegemony in favor of the newly dominant Flemish. Yet we have to remember that the act of cross-breeding that launched the CCP was before all else a material practice in which living beings were brought together to perform the natural act of breeding. How does the scientific aspect of hybridity fit into Vanmechelen's project?

To understand that, we need to turn to his encounter with another scientific discipline: the study of evolutionary biology that had its origin in the work of Charles Darwin.

Evolutionary Biology and Hybridity

The question of hybridity is central to Darwin's *The Origin of Species By Means of Natural Selection or the Preservation of Favored Races in the Struggle for Life* (1859) as well as his later study of agricultural selection, *The Variation of Animals and Plants Under Domestication* (1868), a two-volume study published nine years later. To read *Origin* is to grasp the painstaking attention he paid to the long-term effects of the agricultural practices of inbreeding and outcrossing as well as his growing awareness of geography's role in the origin and development of species—both animal and human. For Darwin, hybridity functioned as a contested boundary marker between varieties, breeds, and even species. It was axiomatic that hybridizing produced sterility. While some naturalists argued from this point that infertility functioned to preserve the integrity of species "in order to prevent the confusion of all organic forms," Darwin took issue with this idea (Ridley 2006, 37–47). How could hybrid infertility possibly be understood as a beneficial variation? And yet if it was not beneficial, what advantage did it confer on the individual or new variety in which it appeared? "Darwin needed to show that hybrid sterility was not an adaptation," Mark Ridley has explained, "both to save his own theory and to confound the creationists" (42). In contrast to the naturalists, Darwin held that sterility was merely an incidental result of cross-breeding; "there is no essential difference between varieties and species," the two blurring into each other over time.[18] In fact, he admitted, cross-breeding often produced not sterility but vitality. Hybrids could be impressively healthy specimens, he acknowledged, because they profited from the "slight changes in the conditions of life" that he held "are beneficial to all living beings."[19]

"We see this acted on by farmers and gardeners in their frequent exchanges of seed, tubers, etc., from one soil or climate to another, and back again. . . . Again, both with plants and animals, there is the clearest evidence that a cross between individuals of the same species, which differ to a certain extent, gives vigour and fertility to the offspring; and that close interbreeding continued during several generations between the nearest relations, *if kept under the same conditions of life*, almost always leads to decreased size, weakness, or sterility" (Darwin 1859, 222–223; my italics).

In *Origin* Darwin studied the impact of hybridity upon a species' morphology and fertility, drawing on naturalist William Bernhard Tegetmeier's important work with poultry. He argued that there was a sole parent to the entire species of *Gallus gallus*. "Having kept nearly all the English breeds of the fowl alive, having bred and crossed them, and examined their skeletons, it appears to me almost certain that all are the descendants of the wild Indian fowl, *Gallus bankiva*" (Darwin 1859, 22–23).[20]

By the time of *Variation*, Darwin was still struggling to tease out the distinction between variety and species. *Variation* documents in detail the forms of selection at play in the processes of agricultural domestication in a scrupulous survey of the changes Darwin was able to "collect or observe" in a large number of domesticated species ranging from domestic dogs and cats, horses and asses, pigs, cattle, sheep, goats and rabbits, to pigeons, fowls, goldfish, hive bees and silk moths, and finally to plants—both ornamental and cultivated— flowers, trees, and "culinary plants." (Browne 2002, 285; and Darwin 1868, 170–201). Although the volume met a lukewarm response, selling only five thousand copies and being criticized by reviewers as "excessively detailed," it is well worth returning to for its attention to the relations between agriculture and human culture (Davis and Morris 2007, 411).[21] Selection resulting from domestication need not be intentional to produce dramatic variations in a species, Darwin explains.[22] Whether methodical, unconscious, or natural, selection during domestication both relies on and produces the variability that characterizes organic life. Darwin observes that the two reproductive methods employed in selection have different results: inbreeding enfeebles a breed, while hybridizing frequently "adds to the size, vigor, and fertility of the offspring" (Darwin 1868, 2:153). Under domestication both strategies were in common use, and Darwin gives a number of examples of their effects. Among them is the case of Sir John Sebright's "famous Sebright Bantams," whose conformation and fertility were damaged by continued close inbreeding, so that they "became long in the legs, small in the body, and bad breeders. He produced the famous Sebright Bantams by complicated crosses and by breeding in-and-in [that is, by breeding two generations of siblings to each other]; and since his time there has been much close interbreeding with these animals, and now they are notoriously bad breeders" (2:101). It seemed probable, Darwin went on to argue, that ancient peoples had domesticated not one but several species of chickens, "all being now either unknown or extinct, though the parent-form of no other domesticated bird has been lost."[23] While domestic breeds of chickens have been crossed so frequently their fertility is clearly robust, Darwin uncomfortably acknowledged, "the four known species of Gallus when crossed with each other, or when crossed, with the exception of *G. bankiva*, with the domestic fowl, produce infertile hybrids" (1:246). Yet he scoffed at the notion that some contemporary domestic breeds have emerged from other species of fowl now extinct, concluding whether the method used to produce them is close inbreeding or crossing (hybridization), chicken breeds seem to have "diverged by independent and different roads from a single type" (1:232).

Despite the dispute between "single creationists," those who believed that each species was developed individually at the moment of divine creation, and those who believed that all of the species had resulted by gradual development from one original species, he refused to reason simply from authority when considering the origin of the domestic chicken (1:238). "Authority on such a point goes for little," he argued, declaring instead that he would counter the single creationists by substituting reasoned argument for received authority and providing that all chickens evolved from one original breed. Despite reports by "one ancient author, [who] speaks of fowls as having inhabited S. America at the period of discovery," Darwin declares, "it is . . . as improbable that *Gallas* [sic] should inhabit South America as that a humming-bird should be found in the Old World" (Darwin 1868, 1:245).[24]

Darwin's interest in proving a single origin for all chickens was not merely an ornithological obsession. He was writing in an era when it was accepted that "domestic chickens lay at the far end of an attenuated analogical thread stretching all the way up to MAN. What went for poultry went for God's chosen being" (Desmond and Moore 2009, 219).[25] Darwin's insight into the sole origin of the domestic chicken would receive extended treatment in *The Descent of Man and Selection in Relation to Sex*, where reasoning from the animal to the human he pressed "the main conclusion arrived at in this work," which he knew was sure to meet strong opposition, "that man is descended from some lowly organized form" by way of another conclusion he assumed to be unquestionable: "there can hardly be a doubt that we are descended from barbarians" (Darwin 1871, 919).

In fact, the debate over the meaning of hybridity in the barnyard, so central to Darwin's argument in *Origin*, must be put in the broader context of the emerging scholarship on the origins of the human races, which by the 1840s and 1850s had made hybridity a highly charged term (Young 1995; and Desmond and Moore 2009). A fierce debate between monogenists and polygenists had sprung up. These terms were anything but value-neutral: saddled with the outdated taint of religiosity, monogenists were associated with a fanciful belief in an Edenic origin, while polygenists had the aura of fearless scientific investigation, both archaeological and geological.[26] In the monogenist/polygenist dispute, those who favored a view of the human races as separate species held them to be the product of multiple creations, building their argument on the evidence of agriculture, where, as we have seen, hybridizing, or crossing, often produced partial or complete infertility. One such thinker was one of the most notable poultry breeders of the nineteenth century, the Reverend Edmund Saul Dixon, author of *Ornamental and Domestic Poultry*, who in an article in the *Quarterly Review* attacked Edward Blyth for having a "thirst after 'origins,' that is, *common* origins in domestic fowl" (Desmond and Moore 2009, 219). The question of multiple origins was so important, Dixon explained, because it proved that breeds could not be changed either by agricultural domestication or by environment. Instead, he argued that the "main domestic races are each derived from a distinct species, in dogs, in pigeons, and in *people*" (219).

As a broader theoretical concept, hybridity anchored fields as diverse as linguistics and racial theory, where, in the form of pidgin languages and children born of miscegenation, it marked two major forms of contact between the colonizing North and the colonized South: language and sexuality (Young 1995). From its nineteenth-century roots in the project of racial taxonomizing and the regulation of colonial desire, investigations of hybridity in race science converged with investigations of cultural supremacy: "The cultural question was whether there had ever been a black civilization: if not, this would substantiate claims about the superiority of the white race and the inferiority of the black; the biological question was whether the hybrid offspring of unions between the two races were fertile or not: if not, this would show that they were different species" (124). By the twentieth century, hybridity seemed to have shifted to the realms of culture and society, where it motivated explorations of the role of discourse and desire in the production and maintenance of racial and cultural hierarchies (Young 1995). Yet, in reality, the old race science was still using hybridity to consider questions of cultural and racial supremacy: the site of that investigation had just shifted from the human realm to the animal.

The Debate over Breed Origin

In 2001, investigation of contact between civilizations and races in the human species came together with the scientific studies of breed origins and varieties in the avian species. The analysis of the mitochondrial genome of the chicken was complete by 2001 and the complete sequencing of the chicken genome (*Gallus gallus*) was accomplished in 2004 (Storey et al. 2007, 10339; see also Wong et al. 2004). Armed with methods and technologies unavailable in Darwin's pregenetic era, avian reproductive biologists, anthropologists, and geneticists again took on the issue of species origins, providing some stimulating challenges to Darwin's assertion of a single origin for the domestic chicken. Returning to Darwin's questions, researchers in fields ranging from molecular biochemistry, animal breeding and genetics, to archaeology, molecular ecology and evolution now frame them very differently. Often they articulate them in terms that testify to the intimate links between biological and cultural hybridity or, to put it in the terms used, between reproductive fertility and cultural hegemony. Was the South American chicken descended from a European variety brought by Columbus and the conquistadors in the fifteenth century, or did it reflect pre-Columbian contact between Polynesian people and the native peoples of South America? Was the red jungle fowl the ancestor of both European and South American breeds of chicken? In the culturally inflected return to the question of avian origins that had interested Darwin as a refutation of the

doctrine of a single creation, we will find reworked some of the critical categories of that nineteenth-century notion of hybridity, in particular the debate
between monogenetic and polygenetic explanations of human origin with its
implications for questions of nationality, ethnicity, race (and racialization) and
culture. These issues too provide some of the context for Koen Vanmechelen's
reimagining of Gallus gallus in his Cosmopolitan Chicken Project.

In 2007 Dr. Alice Storey of the department of anthropology and the Allan
Wilson Centre for Molecular Ecology and Evolution of the University of Auck-
land, New Zealand, published a paper with several colleagues that claimed
"firm evidence for the pre-Columbian introduction of chickens to the Ameri-
cas" and strongly suggested that the South American chicken had been introduced by Polynesian voyagers. This finding countered the long-held notion
that chickens had been brought to South America by Spanish conquistadores.
Dr. Storey and her group argued that that a single chicken bone uncovered during an archeological dig on the Arauco Peninsula, Chile, carbon-dated to be
roughly six hundred years old, possessed a very rare genetic mutation resembling one found in bones uncovered at prehistoric sites in the Pacific: one
at Mele Havea on Tonga, and one at Fatu ma Futi in American Samoa (Storey et
al. 2007). The very next year, Dr. Storey's theory was refuted by an article in *Proceedings of the National Academy of the Sciences* published by Dr. Alan Cooper, of
the University of Adelaide in Australia, and his research group (Gongora et al.
2008; see also GenomeWeb staff 2008). Cooper and colleagues compared the
DNA sequences of forty-one species of native Chilean chickens to an online
database of more than one thousand sequences of domestic chickens existing
worldwide, and advanced an alternative theory of origin for the South American chicken: "We consider the clustering of the modern Chilean chicken
sequences with [groupings] predominant in Europe to indicate the contribution of Spanish-introduced chickens" (Gongora et al. 2008).

Cultural as well as genetic evidence was factored in as Cooper's group challenged Storey's findings. Anthropological understandings of human behavior as
well as naturalist understandings of chicken behavior suggested to Cooper and
colleagues that Dr. Storey's attribution of a pre-Columbian origin to the
chicken bone was incorrect. As they reasoned, chickens, who are scavengers,
will eat any foods available. This settlement was less than three kilometers
away from the ocean; both the large middens, or trash piles, found there and
the fact that the archaeological site was less than three kilometers away from
the ocean suggested that the diet of this chicken had included shellfish. Cooper
and colleagues suggested that the presence of marine carbon had contami-
nated the carbon dating process. In reality, the bone was of much more recent
origin: "If the diet of the El Arenal chicken included a marine carbon contribution of more than 20%, the calibrated age would be post-Columbian"
(Gongora et al. 2008).

Dr. Storey refuted the argument that her data was tainted by the presence of marine carbon, arguing that Dr. Cooper's reliance on contemporary DNA skewed his findings: all chickens will show sections of similar DNA sequences, having all come from Asia. "We would absolutely expect all chickens, everywhere in the world, to have an Indian/Asian genetic signature as all chickens must come from one of those places. However, there was probably more than one region in which chickens were domesticated, like India, China or Thailand. This means that all chickens everywhere in the world have a mtDNA (mitochondrial DNA) signature that relates back to their Asian great, great, . . . grandmothers, and those will differ depending on the domestication centre" (D. Cooper 2008).

The rhetoric of each group betrays a broader clash of disciplinary worldviews. Cooper's group grasped what would make this contemporary debate newsworthy: not the question of the origin of the South American chicken per se but rather the evidence it might offer of contact between indigenous South American peoples and Polynesian peoples before Columbus came to the new world. Yet the language of Storey's refutation recalls Darwin's argument in *Variations*—the importance of domestication not only as a cultural practice but as a crucial shaper of biological materiality: "The maternal lineages [mtDNA] of domesticated animals have been manipulated by breeders for many years—so modern chickens are a mixed lot. If you really want to understand what lineages are in a given place at any point in time, you need to study the ancient DNA from dated archaeological remains" (D. Cooper 2008).

In early 2008 a third group of researchers entered the discussion with a dramatic finding. "Darwin Was Wrong about Wild Origin of the Chicken, New Research Shows," reported *Science Daily*, relying on materials furnished by Uppsala University.[27] While these researchers were more restrained in their own publication, which appeared in the online open-access journal *PLoS Genetics*, they also emphasized the advance over Darwin that their research accomplished: "On the basis of observed character differences and cross-breeding experiments, Darwin concluded that domestic chickens were derived solely from the red junglefowl, though this was later challenged by Hutt, who stated that as many as four different species of junglefowls may have contributed to chicken domestication" (Eriksson 2008, 2).

Moving beyond Darwin's attention to poultry behavior and cross-breeding, and beyond twentieth-century researchers' study of mitochondrial DNA and retrovirus sequences, all of which yielded support for the thesis of a single origin for the domestic chicken, this new study turned to gene sequencing technologies to challenge Darwin's thesis. Skin color in birds varies between white, black, and yellow, with domestic chickens generally being either yellow skinned (only those birds whose parents both carried the specific recessive gene for yellow skin color) or white skinned (those birds with at least one parent carrying the dominant allele). The yellow color in a bird's skin comes from beta-carotene in the food it eats, which in birds with a specific pair of inherited genetic mutations will produce skin with a yellow color. The accomplishment of Eriksson and colleagues was to demonstrate that birds inheriting two copies of a mutant regulatory gene will have yellow skin whereas those with at least one copy of the nonmutant regulatory gene will be white-skinned because the gene will have turned "colorful carotenoids into colorless apocarotenoids" (Eriksson et al. 2008, 2).

While Darwin observed plumage color at the level of the organism and Eriksson and colleagues studied skin color at a molecular level, their findings both demonstrate the persistent connection between color and hybridity. Darwin described the crosses he made: "I first killed all my own poultry, no others living near my house, and then procured, by Mr. Tegetmeier's assistance, a first-rate black Spanish cock, and hens of the following pure breeds,—white Game, white Cochin, silver-spangled Polish, silver-spangled Hamburgh, silver-pencilled Hamburgh, and white Silk. . . . Of the many chickens reared from the above six crosses the majority were black, . . . some were white, and a very few were mottled black and white" (Darwin 1868, 1:247–248). Eriksson and colleagues went Darwin one better: they demonstrated that there were two genes for skin color in the domestic chicken. Because the genes for skin color they identified are hereditary, they explain, their different proportions in domestic chickens suggests that our current barnyard chicken has at least two different ancestral origins: "though the white skin allele originates from the red junglefowl, the yellow skin allele originates from a different species, most likely the grey junglefowl" (Eriksson et al. 2008, 2). Moreover, they argued, the hybridization between those two original chickens was most likely the result of agricultural intervention rather than natural selection.[28] The layman's abstract accompanying their article in *PLoS Genetics* mingles a consideration of the specifically scientific implications of this finding with discussions of the agricultural implications of skin color in chickens.[29]

> The majority of chickens used for commercial egg and meat production in the Western world are homozygous for the yellow skin allele. . . . There is a strong consumer preference for the yellow skin phenotype in certain geographic markets such as USA, Mexico, and China where synthetic pigment may be added to enhance the yellow color. . . . It is therefore worth speculating that the bright yellow skin color, expressed by well-fed yellow skin homozygotes but not by well-fed white skin birds, has been associated with high production and good health at some point during domestication and was therefore favored by early farmers. Of course, yellow skin may also have been selected purely for cosmetic reasons.
>
> (Eriksson et al. 2008, 1, 4)

"Culture and history must be rethought with an understanding of their inextricable, if highly variable, relation to biology," argue Lennard J. Davis and David B. Morris (2007, 411). By incorporating geographically variable preferences for skin color in chickens with global farming and marketing strategies, these three scientific studies return to Darwin's question of the single origin of the domestic chicken and offer updated perspectives on the densely interwoven biocultural relations linking domesticated animals and the human cultures that raised them. Considered in light of these recent scientific investigations of the origin of the domestic chicken and the race science subtending and following from that project, which is now submerged into the discourse of economic and cultural globalization, Vanmechelen's process of breeding toward his hybrid super-bastard chicken raises some troubling questions, just as it did when we considered it in light of the evolving meanings of cosmopolitanism and hybridity.

From Vanmechelen's first cross-breeding of the Mechelse Koekoek and the Poulet de Bresse, each act of hybridizing is also an act of cultural assimilation—an act that crosses a pure breed from another nation with a hybrid whose name carries on the poultry patrilineage of the most local privileged breed, the Mechelse. Moreover, if we examine photographs of the succession of Vanmechelen's cosmopolitan chickens, looking not just with the agricultural eye of one familiar with poultry breeding but also with linguistics and nomenclature in mind, it is clear that each act of hybridizing preserves even as it eradicates. Although the purebred chicken is transformed into a hybrid, with the illegitimate lineage its bastard label implies, the lineage of one breed—the Mechelse—is still maintained. Thus the male Mechelse Koekoek and the French Poulet de Bresse produce the Mechelse Bresse; the male Mechelse Bresse crossed with the English Redcap produces the Mechelse Redcap; the male Mechelse Redcap and the American Jersey Giant produce the Mechelse Giant; the male Mechelse Giant and the German Dresdner produce the Mechelse Dresdner; the male Mechelse Dresdner and the Dutch Owlbeard becomes the Mechelse Owlbeard; the male Mechelse Owlbeard crossed with the Louisiana from Mexico becomes the Mechelse Louisiana; the male Mechelse Louisiana crossed with the Thai Fighter becomes the Mechelse Fighter; the male Mechelse Fighter crossed with the Auracana from Brazil becomes the Mechelse Auracana; the male Mechelse Auracana crossed with the Turkish Denizili Longcrower becomes the Mechelse Longcrower, and so on.

If we extrapolate from Beck and Sznaider's notion of cosmopolitanism as a method of inquiry to include the case of Koen Vanmechelen's art, we can argue that it should call into question all oppressive binaries: of cartography, nationality, sexuality, and species. Yet Vanmechelen's experiment with the cosmopolitan chicken ends up reinforcing the identity of the author himself: a Flemish man, in implicit comparison with every other nationality and ethnicity. And here, perhaps, he makes his most disturbing return to Darwin, not to his scientific vision but to his scientific racism. For in the end, his breeding project to affirm the "hybrid super-bastard chicken" recalls the act of rhetorical recentering of the European subject by way of transspecies identification with which Darwin ends his exploration into human and animal breeding, *The Descent of Man*: "For my own part I would as soon be descended from that heroic little monkey, who braved his dreaded enemy in order to save the life of his keeper, or from that old baboon, who descending from the mountains, carried away in triumph his young comrade from a crowd of astonished dogs—as from a savage who delights to torture his enemies, offers up bloody sacrifices, practices infanticide without remorse, treats his wives like slaves, knows no decency, and is haunted by the grossest superstitions" (Darwin 1871, 404–405).

In his conclusion to *The Descent of Man* Darwin invokes the metaphoric animal to cement its, and by extension his, superiority over the specific, material human being—the "savage"—who is marked in contrast by racial and cultural otherness. In a similar rhetorical move in Koen Vanmechelen's Cosmopolitan Chicken Project, material chickens existing on their own terms have been subsumed into an ideal, metaphoric chicken whose terms of existence are purely human.

A closing anecdote can sum up the complex position of the hybrid in chicken culture as well as bring us to the threshold of the final aspect of chicken culture we will explore in "Inauguration." In an article for the *Guardian online*, Mike Phillips, the author of *London Crossings: A Shadow of Black Britain* and *A Shadow of Myself*, evoked the impact of Barack Obama's victory in the U.S. presidential election of 2008 for those outside the United States:

> I was in Belgium on Tuesday night, having dinner at the studio of the artist Koen Vanmechelen with a group of art collectors. We were talking about an exhibition I'm going to curate in Venice next year and no one mentioned Obama during the dinner, but he was never far from my thoughts. We were talking about budgets and contracts and my attention was concentrated, but the fact that part of my mind was on Obama gave me a strange new confidence to say what I wanted and how I wanted things to be.
>
> On the way back to his house, Koen, a Belgian, talked about hybridity, comparing the freedom with which he roamed around the world's cultures with Obama's position. I didn't argue, partly because I was marveling at the sheer ease with which it seemed possible for almost everyone I was encountering in Europe to identify themselves with the man. When we said goodbye, Koen gave me a thumbs-up: "Obama."

(Phillips 2008)

In its cosmopolitan wryness, Phillips's anecdote testifies to the contradictions that make Vanmechelen's art interesting and troubling. As Phillips implies, Vanmechelen's cosmopolitanism may be very different than Obama's for reasons that transcend the author's good intentions, reflecting the uneven and oppressive hierarchies of race, class, sex, and culture that the concepts of hybridity and cosmopolitanism have been used to measure and enforce.

Inauguration

"Some people may call this theft."

–Booker T. Washington, *Up from Slavery*

Writing before the swearing-in ceremony for the forty-fourth president of the United States, and the first African American president, *Washington Post* reporter John Kelly recalled the original meaning of the term *inauguration*:

> The upcoming inauguration of Barack Obama allows us to have a quick lesson in Roman history. Though its precise etymology is disputed, "inauguration" is generally believed to come from the Latin for "directing the birds." Birds were very important to the Romans for foretelling the future and discerning the moods of the gods: the way birds flew, what they ate, how they ate. (It was considered a good sign if food dropped from the beaks of the sacred chickens as they were fed. Yes, there were sacred chickens).   (Kelly 2008)

Rather than recalling Roman history, we would do better to look to American history if we want to consider the role of chickens—sacred or profane—in the historic candidacy of Barack Obama. In particular, we should review the history of racial stereotyping in the United States dating back to the era of slavery.[1] And, because as we know chickens are good to think with, we should use them to think about the intertwined relations of race, law, and property embodied in the image of the black chicken thief.

Historian Psyche Williams-Forson has traced the "perverse and overwhelming" archive of images "of African American men . . . in compromising positions with chickens" that testifies to a longstanding association between black men and chickens (Williams-Forson 2006, 38).[2] She holds that in addition to expressing the genuine affinity among African Americans for a relatively inexpensive food source that has long provided nourishment, a route to self-determination, and a source of agency for blacks in a racist, inequitable society, the connection between black men and chickens also served a number of more complex and disturbing functions in the antebellum period

(Williams-Forson 2006). It satisfied the white need to recall a prewar era of black slavery and white racial dominance; served whites as a humorous release from the grinding need to maintain control over a newly freed black population; testified to the unease whites felt at the economic resilience and resourcefulness of the propertyless black population; embodied the intimidation white people experienced at the fantasized greater sexual prowess of black men, those so-called bucks with large "cocks" and an insatiable sexual drive; and naturalized the negative traits associated with the shiftless, lazy, unreliable chicken-stealing black man.

The image of the chicken was also deployed, I would argue, to express (and defuse) the anxiety whites felt every time an African American "first" took down another racial barrier. Thus two iconic sporting "firsts," boxing heavyweight champions Jack Johnson and Joe Louis, were repeatedly portrayed as gaudily dressed, black dialect–speaking Sambos slurping up watermelons, gorging on chicken, and hiding from work. This "chicken and watermelon eater" racial stereotype, as historian Lawrence D. Reddick describes it, took an even more pointed turn when Joe Louis won the heavyweight boxing crown in 1938 with a first-round knockout of Max Schmeling. A cartoon in a New York newspaper portrayed him "as a chicken thief trapped in farmer Max Schmeling's henhouse. As the angry, rifle-toting Schmeling approaches the chicken coop, the cartoonist warns Sambo Louis with this caption, 'Look Out, Joe, Here Comes a Good Shot'"(Wiggins 1988, 249).[3]

Ten years before Joe Louis's famous victory, a toy marketed in the Sears, Roebuck and Company catalogue for 1928–1929 suggests the stereotype was as common as child's play. As the catalogue description emphasizes, the new toy, named the "Chicken Snatcher," was designed to appeal through its kinetic portrayal of the black thief's humiliation and fear at being caught stealing a chicken. "One of our most novel toys. When the strong spring motor is wound up the scared looking negro shuffles along with a chicken dangling in his hand and a dog hanging on the seat of his pants. Very funny action toy which will delight the kiddies" (Nelson 2001).[4] The black man with his purloined chicken and watermelon red waistcoat reappear in a slightly different form in a stereoscopic photograph published by Underwood & Underwood of New York.[5] In this image, which also elaborates on the racialized drama implied when a black man and a chicken are pictured together, an elderly, gray-bearded black man in bowler hat, jacket, and waistcoat over torn and patched trousers stands in a farmyard holding a large striped watermelon under each arm. He is looking down at a large white and gray chicken. The caption reads, "Jes' dis Niggah's fool luck!—bofe arms full an' dat rooster a beggin' to be took along" (Williams-Forson 2006, 54).

These two representations, combining the elements of black man, watermelon, and chicken, anticipate one of the most notorious images to circulate in

FIG. 23. Obama Bucks

the United States presidential election of 2008: the supposedly humorous "Obama Bucks." This "Food Stamp" dollar (or "buck," to play on the racist stereotype of the hypervirile black man, or "buck") figured fried chicken and watermelon, food items central to the racist stereotype of the black chicken thief, flanking a dancing brown donkey with Barack Obama's face. This chapter will explore how the chicken imagery exemplified in the Obama Bucks operates as a condensed node of social, political, and economic significance surrounding the candidacy of the first African American man for president of the United States.

The Obama Bucks image was created by fervent Obama supporter Timothy Kastelein, associate chair of a Democratic Party precinct in Minnesota, who explained that he had hoped to "satirize the fringes of the Republican Party who fear a black president" (Furst 2008). Kastelein may have attempted to stabilize and sanitize the image's motivations and connotations, but, like all other images, it is both overdetermined and uncontrollable. As a Web image, its circulation could not be confined to one ideological perspective. Diane Fedele, president of the Chaffey Community Republican Women, Federated, included the image in a newsletter she distributed to Republican supporters. When Fedele was forced by the resulting outcry at this racist stereotype to tender her letter of resignation, she explained that she "did not know the food items were African-American stereotypes and were considered insulting. 'I do not think like a bigot,' she wrote" (Furst 2008). Yet whether Fedele was conscious of it or not, the iconography of the "Obama Bucks" image dates back to the eighteenth century, when broadsheets such as the *Colored American*, *Christian Reporter*, and *National Era* reported "articles related to African American chicken husbandry, preachers and their love of chicken and fowl, and concern over African-American chicken stealing" (Williams-Forson 2006, 19). Indeed, several of its iconic images—the watermelon, the bucket of chicken, and the dancing donkey—are central to the complex tradition of racist iconography in the United States.

The Racialization of Theft

Central to the image of the black chicken thief is the act of theft itself, an act that both social custom and law operated to frame along racial lines as one among other "deviant tendencies." *Race, Theft, and Ethics: Property Matters in African American Literature* (2007), feminist scholar Lovalerie King's historically grounded reading of African American literature, argues that racism in the United States has become normalized in the stereotype of the black thief (King 2007, 23). Drawing upon African American literature, which represents a crucial record of struggles over the meaning of property rights in the context of slavery, King explains that because the legal code of the United States reflected the interests of the slave owners, the act of theft became associated with black identity. Blacks were legally forbidden to own property above the necessities to sustain life, and any means of acquiring property was therefore "necessarily defined as outside the law, as theft" (42). However, the act of appropriating property belonging to the slave owner, like a chicken or pig taken from a plantation farm, was understood within the black community not as "stealing" but as "taking," reflecting the greater crimes inherent in the institution of slavery that had been committed by the slave-owning whites from whom the property had been taken: "kidnapping and the plunder of home, family, labor, identity, and history" (30). Two basic and opposed notions of theft are at play in the image of the black thief, according to King: the legal expropriation of black Americans' selfhood and freedom (accomplished by a system of laws propping up slavery and racial discrimination) and the illegal instances in which slave or free blacks took items that were legally the property of those who enslaved or oppressed them (29). Assessing the ethical meaning of the slavery experience and adopting what King calls "the rhetoric of reparations," we are able to understand that the greater act of thievery belongs to the slave owner, for he (or she) has stolen the liberty, kinship ties, work, and very selfhood of his slaves.

Both the photograph of the old black man with the watermelons under his arm and the tempting chicken at his feet and the Obama Bucks image from 2008 reflect this racially linked understanding of property law dating back to the institution of slavery. The slice of watermelon on the Obama Bucks references the watermelons taken from the farm field; the bucket of Kentucky Fried Chicken recalls the chickens lifted from the hen house; the barbecued ribs allude to the squealing purloined pig. Yet while participating in the tradition of stereotyped representations of the black thief, the Obama Bucks image brings it up to date by incorporating the fears of contemporary conservative voters (as well as, arguably, the racism of contemporary liberal voters). These "lifted" food items appear on food stamp dollars, part of the contested United States welfare program that are imaged here as theft within a legal system that grants property rights only to racially and economically dominant whites and is dedicated to obscuring the far more damning theft of the slave's labor, self-hood, and freedom. Indeed, the racialized food associations link the entire food stamp program to a single race, imaging it as black only.

While the donkey itself has a long history as iconic image of the Democrats, the meaning of the smile on the face of the dancing donkey at the center of the Obama Bucks—Barack Obama's smile with a Cheshire cat set of historical resonances—is as plurivalent as the image as a whole: does it protest stereotyping or confirm it? The smile either reflects a black man's ill-gotten gains (chicken, watermelon, food stamps, and other unmerited assistance measures provided by Big Government) or it attests to Obama's skill at negotiating the complex dance so long required of black Americans in a racist society. Neither political perspective is immune to the persistent legacy of racism or to the image of blackface minstrelsy that provided a major part of the context for the image of the black chicken thief. The minstrel show captured what Eric Lott has called "racialized elements of thought and feeling, tone and impulse, residing at the very edge of semantic availability" (Lott 1992, 6). Understood as a "pre-cognitive form: not, as in Geertz's study of the Balinese cockfight, a story one people told themselves about themselves," to Lott blackface minstrelsy consisted in "the dialectical flickering of racial insult and racial envy," a mixture of "theft"—a racist cultural appropriation (or "expropriation") of black cultural forms—and "love"—a fascination with the black male body itself (Lott 1993, 18, 8). While Lott links blackface to other public displays that according to Victor Turner operate as modes of "public reflexivity" linked to "times of radical social change" as part of the "repertoire of prophetic leaders who mobilize the people against . . . overlords threatening their deep culture," the blackface in the Obama Bucks is a far more complicated performance (Turner 1977, 36). It has recursive significance, for in this instance the mixed-race man behind the dancing donkey has a racial identity that both inverts and subverts the standard white-man-in-blackface identity.

Just as minstrelsy has functioned as "a ground of American racial negotiation and contradiction," so too the Obama Bucks image can be seen as a site of negotiation of the particular meanings that this particular presidential candidate, and his candidacy, held for the American people (Lott 1993, 30). Frederick Douglass saw blackface minstrelsy as "a site of political struggle for representation, debased and suspect though it might have been," arguing that "it is something gained, when the colored man in any form can appear before a white audience" (Douglass 1849). The current reality of an African American as president—and a black family in the White House—rings changes on the old fantasy of what an election might mean to the African American community in the days of slavery and the Reconstruction-era south. From 1750 on, Lott tells us, Election Day consisted in a day of wildly festive inversion in African American communities, when blacks chose their own leaders and exercised a parodic reversal of white rule, white culture, and white law. As the old racist stereotype "Zip Coon," the urban opposite of the plantation "darky," imagined it in the eponymous song by George Willig Jr. (1834), the election of an African American president would reverse the legal standing of whites and blacks.

> If I was de President of dese United States,
> I'd suck lasses candy and swing upon de gates,
> And dose I didn't like I'd block em off de docket,
> And de way I'd block em off would be a sin to Crocket.   (Lott 1993, 179)

Another version takes the narrative even further, linking the protagonist's courting of "old Suky blue skin," who lives on "chicken foot and possum heel, widout any butter" (traditionally foods purloined by those who had little other food source) to not only an economic crash but a transformed presidency:

> I tell you what will happen den, now bery soon,
> De Nited States Bank will be blone to de moon;
> Dare General Jackson, will him lampoon,
> And de bery nex President, will be Zip Coon.[6]

In their mixture of blackface stereotype and intimations of political and social disruption, the Zip Coon songs reflect the complex, conflicted performances that constitute the American black identity, a preoccupation they share with another form of blackface entertainment: the early "race films" (Lott 1993, 37).[7]

Race Films and the Chicken Thief

As I have pointed out, several of the iconic images on the Obama Bucks—the watermelon, the bucket of fried chicken, the dancing donkey—recall the race movies of the early twentieth century. The watermelon references the black man's supposed sexual appetite for its luscious pink flesh; the dancing donkey, like the currency itself, compactly alludes to the "buck dance" films released by several film companies at the turn of the twentieth century, which showed uninhibited, primitive, even grotesque dancing by African American males competing for a watermelon prize; and the bucket of chicken recalls the most widely circulated filmic stereotype according to film historian Gerald R. Butters: "African-American men as chicken thieves" (Butters 2002, 22–24). Like other forms of blackface minstrelsy, race films "flaunted as much as hid the fact of appropriation and its subtexts, enslavement and intermixture" (Lott 1992, 44). Relying on stereotypical images for a large part of their appeal, Butters tells us, race movies were shown in segregated theaters so that white movie goers could laugh without inhibition at the debased blackface caricatures of African Americans (Butters 2002, 7). Yet, like the Obama Bucks image, these race films featured iconic imagery that was notoriously unstable, able to undermine as well as affirm the state of race relations of their era.[8]

The classic example of this iconic imagery was the chicken theft plot, a staple of early race movies. Beginning with the Edison Company's *Chicken Thieves* (1897) and continuing through several versions released by Sigmund Lublin's film company in the early twentieth century and American Mutoscope and Biograph Company's 1904 hit production *The Chicken Thief*, a series of films used black actors to depict the stereotype of the black chicken thief.[9] The plot was standard: the thief was shown sneaking into the chicken house, stealing chickens (more or less successfully depending on the vigilance of the white farmer), and then in many cases joining other men and women for the stereotypical dance (Butters 2002, 25). An advertising bulletin for the 1904 Biograph film *The Chicken Thief* captures the essence of this highly popular (among white audiences) film genre and demonstrates the features it shares with the image of the Obama Bucks. In listing the "People in the Play," who are almost always identified by race, the advertisement gives no racial particulars for the central characters, suggesting that their race is self-evident: "First Chicken Thief, Second Thief" (Niver 1971, 140–143).[10] The plot description is similarly straightforward, promising: "a prologue and four acts," as follows:

PROLOGUE.—Showing the First Chicken Thief in relief.

ACT I.—Interior of Chicken-Coop—Night. Exterior of Chicken-Coop and Barnyard—Next Morning.

ACT II.—Interior of Negro Cabin—A Chicken Dinner.

ACT III.—The Flight Through the Woods. (Film tinted throughout to give moonlight effects.) *In five exciting scenes.*

ACT IV.—Interior of Negro Cabin—The Capture.

(Marion and McCutcheon 1904, 2)

The *Bulletin* goes on, in a lavish scene-by-scene description of the film, to praise *The Chicken Thief* for its "Rip Roaring" humor:

From the opening of the picture, where the coon with the grinning face is seen devouring fried chicken, to the end where he hangs head down from the ceiling, caught by a bear trap on his leg, the film is one continuous shout of laughter. The opening scene is a triumph of photography . . . a moving picture of the interior of a big hen-coop at night, showing over one hundred chickens asleep on the roosts as the thieves enter. Nothing could be more realistic. With careful moves the experienced coons gather in the fluttering and squawking chickens by the armful, and when the [*sic*] their bag is full to overflowing they clear out as silently as they came. (3)

To readers now, what stands out in this *Bulletin* is not the humor or the technical detail but the fact that an act of desperation intended to feed a family is rendered as a comic lark followed by a nostalgically sentimental domestic idyll. After the hen-house caper, the scene shifts to "the interior of a darky cabin" where a black family is sitting down to chicken dinner at a table headed by the "first chicken thief," with "a colored dominie as the guest of honor" (the slightly contemptuous term for this minister reflecting his race) and "two little pickaninnies . . . feasting at an extemporized table made of a box placed on the floor" (3). Poverty is transformed into vicarious pleasure for the viewer: "The savory stew and the coffee pot are steaming on the stove nearby. And the way those darkies get away with that chicken fricassee is something marvelous. It would make the mouth of a confirmed dyspectic water to look at the scene" (4).

The next five scenes, praised in the *Bulletin* as another model of technical innovation but again most remarkable for their linguistic and performative reduction of African American men to woodland varmints, begin with the chicken thieves returning to the hen-house "for another raid on the roost." "Two coons come sneaking in on hands and knees. . . . Cautiously they make their way to the chicken-coop door when Bing! The foremost coon puts his foot in the bear trap, and the huge jaws snap over his ankle" (4). The "two coons" then flee a posse of white farmers armed with shotguns through the woods. Thanks to some skillful work by a colorist, the flight scene takes place in moonlight, "that pale and mysterious light so dear to the heart of the possum-hunter and his compatriot, the chicken thief" (4). Animal and human, hunter and prey mingle in a topsy-turvy jumble of categories in the *Bulletin* description, as the "coon" flees under the possum-hunter's moon, until he takes refuge in the "negro cabin" where a dancing party is underway, hiding (at the dominie's suggestion) in the attic. The film closes with an image of the "unlucky coon" hanging head-first from the attic, his leg caught in the bear trap.

Like the falsely cozy description of the chicken dinner, the description of this man as "unlucky" as he dangles head-down from the cabin rafters, his leg in a bear trap, avoids the systematic nature of poverty, crime, and punishment in this story of theft (4). In a similar sleight-of-hand, in the Lublin catalog another such film, *Dancing for a Chicken,* portrays chicken theft as a function of black rather than white appetite, and black rather than white competitive self-interest. "A party of colored folks are [*sic*] engaged in a dancing contest for a chicken. Everyone knows that coons love chicken and each buck and wench is doing his or her utmost to win the prize. One imagines one can hear them smacking their lips while devouring the chicken that is yet to be won" (Butters 2002, 25).

Before 1905 the race film was one of the rare subgenres of film in which blacks were employed; it offers complex glimpses of the understanding of property, the law, religion, and community in the era of slavery and Reconstruction

South. However, when it started up again after a lull between 1905 and 1909, the chicken thief genre employed not blacks but white actors in blackface.[11] Perhaps as a result of the changed personnel, other differences also characterized the new iteration of the race film genre after 1909.[12] Violence now played a major role, whether in the form of vigilante groups, wealthy college boys or one disgruntled hired hand, as in *A College Chicken* (1910) and *Hen House Hero* (1912), as did sexuality. The theft of the chicken now condensed a wish far more troubling to the white audience—the theft of a woman's sexual favors— in *Rastus' Riotous Ride* (1914), *The Tale of a Chicken* (1914), *C-H-I-C-K-E-N Spells Chicken* (1910), and *Mandy's Chicken Dinner* (1914) (Butters 2002, 27).

Butters finds only one film that questions the notion of black chicken thievery, the "subversive" short film *Chased by Bloodhounds* (1912). In this short film, Mr. Bunny, a rich white chicken breeder, decides to set the dogs on the poor fat black man who—after receiving a gift of the white man's clothing—has stolen one of his chickens. The tracker becomes the tracked, when the bloodhound, tracing the scent of the borrowed clothing, chases Mr. Bunny to exhaustion. The white man is left wracked by nightmares, second-guessing himself and struggling with guilt over his actions. In contrast, the black man is left with not only the clothes given him by the wealthy white chicken breeder but also one of his high-quality chickens (27–28).[13]

Reconsidering Theft in the Context of Slavery

Using blackface minstrelsy to explore the various meanings of chicken theft, these race films recall what Lovalerie King has called the "literary minstrelsy" performed by Booker T. Washington in his magisterial autobiography, *Up from Slavery* (King 2007, 159n53). There, as Washington strove to make his experiences as a black man palatable to white readers in the interests of obtaining education and philanthropic aid for African Americans, he also reframed the theft of a chicken by situating it in the context of slavery as an institution. "One of my earliest recollections is that of my mother cooking a chicken late at night, and awakening her children for the purpose of feeding them. How or where she got it I do not know. I presume, however, it was procured from our new owner's farm. Some people may call this theft. If such a thing were to happen now, I should condemn it as theft myself. But taking place at the time it did, and for the reason that it did, no one could ever make me believe that my mother was guilty of thieving. She was simply a victim of the system of slavery" (Washington 1901, ch. 1; see also King 2007, 59.) Washington offers a tribute to his mother's business acumen, frugality, and courage, drawing on "a most pervasive stereotype from American slavery—the black chicken thief" to ground a narrative of "systemic and systematic deprivations" (King 2007, 59). As Washington tells it, when his mother took a chicken under the cover of darkness or engaged in the practice of "'snatching' (stealing) time to care for her children," just as when Washington himself practiced "'stealing' time by manipulating the clock at work so that he could make it to the evening school on time," these acts were part of a strategy necessarily adopted under "the system of slavery" to obtain nourishment, whether actual, emotional, or educational (59–60). Washington suggests that within the unjust system of slavery it is just to view his mother's theft of a chicken as the legitimate appropriation of what should rightfully be hers. In contrast, the stereotype of the black chicken thief racializes ethical and moral depravity, implicitly affirming "the connection between whiteness and property while gradually curtailing the rights of slaves or freed Blacks to own property—whether tangible or intangible" (26).

The Chicken Thief as Index to Equal Access

When approached in the context of slavery, with its peculiar definition of property, the image of the black chicken thief calls into question the purported impartiality of the American legal system, particularly the notions of due process and equal protection under the law enshrined in the Fourteenth Amendment of the U.S. Constitution: "All persons born or naturalized in the United States, and subject to the jurisdiction thereof, are citizens of the United States and of the State wherein they reside. No State shall make or enforce any law which shall abridge the privileges or immunities of citizens of the United States; nor shall any State deprive any person of life, liberty, or property, without due process of law; nor deny to any person within its jurisdiction the equal protection of the laws."[14] Analysis of the writings of Frederick Douglass, Harriet Jacobs, Booker T. Washington, and Richard Wright reveals that the very meaning of theft was shaped as an act of resistance to the cultural, social, and legal oppressions arising from slavery and racial discrimination.

Analyzing the role of race in the formation of the American legal process, Lovalerie King recalls two questions raised by A. Leon Higginbotham: "Did the law merely perpetuate old biases and prejudices? Or had it been an instrument first in establishing and only later in attacking injustices based on color?" (King 2007, 163n13). Such questions probe the core of constitutional law, the subject taught by senior lecturer Barack Obama at the University of Chicago Law School. As Jodi Kantor reported for the *New York Times*, "at a school where economic analysis was all the rage, [Obama] taught rights, race and gender." A link to the syllabus for his most popular course, the political and historical seminar titled "Current Issues in Racism and the Law," offered in 2004, corroborates Kantor's observation that "Mr. Obama improvised his own textbook, including classic cases like *Brown v. Board of Education*, and essays by Frederick Douglass, W.E.B. Du Bois, the Rev. Dr. Martin Luther King Jr. and Malcolm X, as well as conservative thinkers like Robert H. Bork" (Kantor 2008).[15] Moving through discussions of slavery, Reconstruction and Jim Crow, civil rights and the backlash against civil rights legislation, Obama's seminar required students do group presentations, choosing between such issues as discriminatory sentencing, criminal laws aimed at specific minorities, criminal profiling (or "statistical discrimination" as it is also termed in the syllabus), and hate crimes.

Copies of the final examinations and examination memos for Professor Obama's Constitutional Law III classes from 2001 to 2003 reveal the context he provided for understanding the relations between race, theft, and the law.[16] This is also the context that can illuminate those stereotypical images adorning the Obama Bucks. As is typical of the genre, the examination questions for Constitutional Law III relied on fictitious case-based questions as routes in to the same substantive issues addressed in the Fourteenth Amendment. The contextual range among these questions is remarkable; they probe the notions of equal protection in relation to race, sexual preference, ethnicity, and nationality in terms of medical treatment, reproduction, social classification and profiling, and finance.

In one instance, a Constitutional Law III question composed after the shock of the World Trade Center and Pentagon attacks concerns a second wave of terrorist attacks involving the deliberate release of a fictitious chemical toxin, rioxin, which can effectively be treated with an antibiotic (Curasin) and a vaccine developed by the same company.[17] Students taking the 2001 final examination are told that that there is evidence that African Americans and Latinos are 15 percent more likely to die of rioxin exposure, and are asked to prepare a memorandum (in the person of assistant to the White House Council) for the president of the United States assessing the constitutionality of a protocol for dispensing the vaccine. "Specifically the President wants to know whether any of the provisions of the proposed protocol violate the Equal Protection Clause of the Fourteenth Amendment."[18]

In the 2003 iteration of Obama's Constitutional Law III class, the final examination features questions introducing a range of other civil rights issues: gay adoption, a voter initiative designed to prohibit "classifying any individual by race, ethnicity, color or national origin" including the practices of "inquiring, profiling, or collecting such data on government forms," and a "high-tech version of traditional 'red-lining' practices, in which banks and insurance companies drew lines around certain minority neighborhoods on a map and instructed their lending officers to avoid issuing credit to businesses and individuals residing in those neighborhoods."[19] All of these fictitious cases address violations of the equal protection clause of the Fourteenth Amendment of the United States. They thus expose as hollow the very same constitutional pledge to dispense even-handed justice that Booker T. Washington challenged in his autobiography.

Like these links to his constitutional law classes, online discussions of Obama's teaching during his presidential campaign provided both a window onto his perspective as a legal scholar and professor and a reminder of the racialized nature of the law. Writing for *The Volokh Conspiracy*, a conservative-to-libertarian collaborative blog, George Mason University Law School professor David Bernstein invited Obama's former students and law school classmates to address the question: "What kind of a Constitutional Law professor was Barack Obama?" Was he "a uniter (open-minded, encouraging critical questions) or a divider (not)? Overtly ideological or dispassionate and neutral? Well-prepared or lackadaisical? Any hints about his constitutional philosophy?" While many of those who responded were concerned with minutiae such as Obama's exact title (senior lecturer), his publishing record (nonexistent), and his ability as a teacher (mixed reviews), one writer addressed more substantive issues. In that post, "B-rob" suggests that Obama's class took an intersectional perspective on the law, incorporating not only race but class, gender, and ability:

> I also had Obama for "Racism and the Law" back in the 92–93 year. . . . I wrote my paper on the "unconstitutional conditions" doctrine as it applied to coercive birth control. How did this relate to "racism" and the law? Because when you trace the history of governmental involvement in forced sterilization, eugenics (*Buck v. Bell*'s "four generations of imbeciles"), etc. (including the castrated chicken thief), you see a very class specific and racially oriented program of (indeed, fascination with) governmental control over poor peoples' and Black peoples' reproduction.
>
> (Bernstein 2008)

"B-rob" reminds readers of the fact that U.S. constitutional law has roots in the institution of slavery and other oppressive institutions of the nineteenth and early twentieth centuries. Central to this post is the same issue that prompted Higginbotham's question: Did the law perpetuate longstanding biases, or did it actually *create* them and only later work to redress them?

In addressing the difference between the mere legal perpetuation of inequality and its legal production, "B-rob" juxtaposes two populations historically subjected to eugenic intervention and governmentally sponsored biomedical control: the poor and the nonwhite. Recalling that he/she wrote a paper on coercive birth control for Obama's seminar, "B-rob" explains that the topic relates to "racism and the law" because it involves the history of governmental sponsorship of eugenic sterilization in the United States, most notoriously exemplified by the 1927 Supreme Court case of *Buck v. Bell*, which upheld the constitutionality of so-called eugenical sterilization of the mentally ill as well as the case of "the castrated chicken thief," *Skinner v. The State of Oklahoma*. By following the trajectory from *Buck v. Bell* (1927) to *Skinner v. Oklahoma* (1942), we can further illuminate the meanings attributed to chickens, blacks, and chicken thieves that formed the context for the cultural response to the presidential candidacy—and, as we will see, the presidency—of Barack Obama.

The Eugenic Roots of U.S. Law: Carrie Buck and James Skinner

A slippage between the terms "race" and "the race" has existed in Anglo-American culture since the late nineteenth century: the former is used to refer to a specific racial group (e.g., blacks), while the latter is used to speak generally of the entire human race.[20] The *Buck v. Bell* case not only embodied the appeal of eugenic discourse in the early decades of the twentieth century as a means of improving "the race" but also provided a stunning example of the law's instrumental power in creating and maintaining injustice (Burgdorf and Burgdorf 1977, 995–996, 1006). With the help of the ringing phrase in Justice Oliver Wendell Holmes's brief, three and one-half page opinion—"Three generations of imbeciles are enough"—this case attained notoriety enough to catapult it into literature and film, where it was memorialized in Abby Mann's story and screenplay and Stanley Kramer's Oscar-winning 1961 film, *Judgment at Nuremburg* (Lombardo 1985; and Kramer 1951). As Burgdorf and Burgdorf report, "One commentator declared that 'the opinion is noteworthy for the boldness with which Justice Holmes dispenses with rules of logic and in short, pithy sentences agrees to the subordination of human rights to the supposed expediency of a long range racial improvement'" (Burgdorf and Burgdorf 1977, 1006).

The *Buck v. Bell* case concerned Dr. Albert Priddy's application to carry out the "therapeutic sterilization" of seventeen-year-old Carrie Buck, a young girl who had given birth to an illegitimate child before being committed—as a "moral imbecile"—to the Virginia State Colony for Epileptics and the Feeble-minded (Kevles 1985). As the supervisor of the colony, Priddy was interested in economizing on patient care, and he intended this as a test case for his policy of preventing the reproduction of patients in order to keep costs down. As prosecuted, this case was a classic example of institutional bias because there were extensive personal ties among the prosecuting attorney, Irving White-head, the defense attorney, Aubrey Strode, and colony supervisor, Dr. Albert Priddy. Moreover, all three men shared the common commitment to the eugenic ideal.

The case record demonstrates a legal system functioning not to redress but to uphold and consolidate inequities of ability, class, and race (in both the broad and the narrow senses of the term). As they were working up the case for the defense, Virginia officials went to one of the most outspoken and passionate defenders of eugenic sterilization, Charles Davenport's right-hand man and fellow chicken-breeder, Harry H. Laughlin, superintendent of the Eugenics Record Office in Cold Spring Harbor, New York. Laughlin was contacted on September 30, 1924, by lawyer Aubrey Strode to request his assistance in "making up a proper record for the test case." What such a proper record might be was in no doubt: as Strode explained in his letter to Laughlin, "Both in the preparation of the statute and in preparation for the trial of this case we have found your book on 'Eugenical Sterilization in the United States,' published by the Psychopathic Laboratory of the Municipal Court in Chicago in December 1922 very helpful" (Strode 1925).

In an appalling miscarriage of justice, Laughlin was called as an expert witness in the case. He testified to "'facts' [that] were direct quotations from Priddy, supplied to Laughlin by mail to inform his eugenical analysis of the Buck family" (Lombardo 1985, 51 n.113). Historian Paul A. Lombardo describes the *Buck v. Bell* decision as a "milestone" not only because it decisively affirmed the government's ability to exercise power over individual rights but also because it endorsed the state's use of biomedicine to serve the ends of government. The law abandoned its duty to the individual in order to enable its use of a practice—eugenic sterilization—endorsed by those in power, based on an uninterrogated stereotype that linked pregnancy out of wedlock to poverty, disease, crime, and social unrest. Laughlin himself would go on to receive an honorary doctorate in medicine from the University of Heidelberg, which he received as "evidence of a common understanding of German and American scientists of the nature of eugenics" (62).

This overarching interest in controlling reproduction for the betterment of the race, whether motivated by the goal of preventing the inheritance of mental illness, physical disability, or so-called inherited criminality, might be said to begin in the United States with the Eugenics Record Office (ERO). It was to the ERO that the defense in the *Buck v. Bell* case appealed, calling on the zealous record keeping, accumulated knowledge, and social interventions of Harry H. Laughlin and Charles Davenport. The ERO pursued eugenic improvement of the race by medical, social, and legal means through agricultural research such as the 1905 paper "Inheritance in Poultry" that Davenport proposed to give at the Lincoln, Nebraska, meeting of the *American Breeder's Association;* through the breeding work carried out at the nearby Station for Experimental Evolution; and through the statistical and social work of the Eugenics Record Office in Cold Spring Harbor, New York. Field workers received eugenic training through traveling to King's Park State Hospital (a psychiatric hospital on Long Island) to hear lectures on "the principle clinical types of insanity"; to Ellis Island to learn of the "clinical symptoms and conditions present in certain types of would-be immigrants"; to Manhattan Eye and Ear Hospital for a lecture by Dr. Walter B. Weidler on "hereditary eye defect, and a clinical demonstration . . . of the particular types of hereditary blindness and defective vision"; to the Coney Island Dreamland Circus Sideshows to hold "impromptu clinics" at the stalls of various "freaks, particularly the dwarfs, giants, and microcephalic idiots"; and finally to the New York Hospital for the Relief of Ruptured and Crippled where they learned about "the hereditary aspect of certain types of human handicap," according to the *Eugenical News* (1921).[21]

The investigations of race, ability, and health carried out by the ERO were grounded in experiments in animal breeding, as dramatized in an illustration accompanying one 1907 article describing the mission of the ERO. The double-page spread in *The World* pictures Charles Davenport, "The Man of Mystery Who Is Searching for the Secret of Life," staring through a large magnifying glass at a newly hatched chick. Symmetrically opposed to him in the illustration, and facing his way, is an authoritatively large Barred Rock rooster, whose leg is banded as indicated by an arrow pointing to it, and a caption reading: "Attached to the Leg of Every Hen and Rooster is a Numbered Brass Tag for Identification Purposes." The article explains the close connection forged by a mutual interest in chicken breeding that had brought together Charles Davenport and Harry K. Laughlin (as we saw in chapter 2), then became the attempt to discover "the secret of life" (as *The World* phrased it) by exploring how the study of animal heredity could illuminate human heredity.

From an early interest in such tangible expressions of heredity as eye, hair, and skin color, Davenport soon branched out to studying human traits that were less subject to measurement such as intelligence and artistic talent. The accumulated "Trait Files" that remain in the ERO records collection contain data ranging from measurable physical malformations to information about individuals in a wide range of professions.[22] Race was a major preoccupation of those gathering trait information, as is evident toward the end of Series I of the Trait Files (A97 #1 through A 9861), where traits of individual races are followed by traits produced by specific race crosses ("Caucasian x Mongolian," "Caucasian x Negro").[23]

When the ERO merged with the Station for Experimental Evolution in 1920 and then became the Department of Genetics at the Carnegie Institution under Davenport's direction, the connection between human genetic inheritance and the multiple fields of animal science fell into the background while the explicit involvement of a wide range of human sciences was emphasized and the explicit interest in race and the effect of racial mixing continued.[24] Yet just as old thought styles persist into new eras, as Ludwik Fleck has shown, so the new community forged around eugenics commitments contained many individuals who had once been, or were still, active in poultry breeding and other forms of applied animal genetics.

Fifteen years after the law affirmed the eugenic sterilization of Carrie Buck, the United States Supreme Court exhibited "a philosophy almost diametrically opposed to the principles enunciated by Justice Holmes" in its 1942 decision "*in re* the case of *Skinner v. the State* of Oklahoma" (Burgdorf and Burdorf 1977, 1023).

This case also concerned the legal right to intervene in reproduction, thus extending the eugenic aim of improving populations, and it too preserved a remnant of the interest in animal breeding characteristic of the earlier thought style of the founders of the ERO.

The second case begins in 1933, when the Oklahoma state legislature passed a law that extended the reach of eugenic sterilization not only to patients in psychiatric institutions but also to inmates in penal institutions too: "It applied to every variety of eugenic 'defective,' covering 'patients likely to be a public charge, or to be supported by any form of charity,' anyone suffering from hereditary forms of insanity, and 'habitual criminals,' defined as three-time felons" (Lombardo 2008, 221–222). The state senator sponsoring the law was Dr. Louis Ritzhaupt, a surgeon who was past president of the State Medical Association. Like Laughlin, Ritzhaupt was also a chicken breeder and exhibitor, preserving the frequent association of medical, eugenic, and agricultural interests characteristic of public men in the first three decades of the twentieth century in the United States. He ran a chicken ranch whose Rhode Island Reds and Polish bantams were shown at national poultry exhibitions, where he would rub elbows with Davenport and Laughlin of the ERO. He would one day be elected the president of the Oklahoma State Poultry Association (221–222). When an attempt by an inmate of the McAlester State Penitentiary to test the 1933 law failed on legal technicalities, Dr. Ritzhaupt went on to sponsor and pass a second law in 1935, the Oklahoma Habitual Criminal Sterilization Act. This act specified that any habitual criminal "who, having been twice convicted . . . of crimes amounting to felonies involving moral turpitude, is thereafter convicted . . . of a crime involving moral turpitude," should be subject to eugenic sterilization (223).

The incarcerated man who tested this new sterilization law was Jack Skinner, himself a believer in eugenic sterilization, who "left home at the age of fifteen in search of work. . . . lost his left foot in an accident, and, at age nineteen, while enrolled in business college, was convicted of stealing six chickens and sentenced to eleven months of hard labor" (Lombardo 2008, 225). Having been convicted twice of armed robbery, Skinner was subject to mandatory sterilization as a "habitual criminal" in accordance with Oklahoma's Habitual Criminal Sterilization Act.[25] Skinner challenged the act as unconstitutional "by reason of the Fourteenth Amendment." He argued that the law made an exception for criminals convicted of white-collar crime. When the appeal case was heard by the U.S. Supreme Court, Oklahoma attorney general Mac Q. Williamson acknowledged that it was difficult to argue that the law was equitable in its provisions. "Asked by the Justices why crimes like stealing chickens made a convict liable for sterilization while embezzlement did not, Williamson admitted frankly that the law's 'exceptions are very difficult to reconcile'" (229). The court decided for the petitioner, holding that the law failed "to meet the requirements of the equal protection clause of the Fourteenth Amendment" to the U.S. Constitution.[26] This new attention to the role of class bias in the Constitution may have been a welcome one to the prison population, according to legal historian Paul Lombardo: "when chicken breeder Ritzhaupt lost to chicken thief Skinner, the convicts at the McAlester penitentiary celebrated in 'an air of jubilation' at news of the victory in their seven-year test of Oklahoma's law" (232).

Just as in the African American literature of slavery and the Reconstruction South, here too the image of the chicken thief anchors a conflict over the interpretation of property law. Assigned to write the formal legal opinion for the Supreme Court, Justice William O. Douglas distinguished between the chicken thief and the white-collar criminal who embezzles money as he struggled to clarify the Court's reasoning: "A person who enters a chicken coop and steals chickens commits a felony (id. 1719); and he may be sterilized if he is thrice convicted. If, however, he is a bailee of the property [that is, if he is entrusted with keeping someone else's chicken] and fraudulently appropriates it, he is an embezzler [id. 1455]. Hence no matter how habitual his proclivities for embezzlement are and no matter how often his conviction, he may not be sterilized."[27] The line between the crimes of embezzlement and larceny is at stake, the court opined, arguing that such legal distinctions have no meaning in eugenic terms, nor do they affect the question of whether criminal tendencies could be inherited.[28] Concurring with the ruling, Chief Justice Stone distinguished between certain forms of mental deficiency, which have been shown by science to be heritable, and the case of criminal tendencies, which "the State does not contend . . . either common knowledge or experience, or scientific investigation, has given assurances that the criminal tendencies of any class of habitual offenders are universally or even generally inheritable."[29] While not challenging the fundamental notion that it is within the rights of the state to prevent those inherited traits that are "injurious to society," Stone observed: "A law which condemns . . . all the individuals of a class . . . because some or even many merit condemnation, is lacking in the first principles of due process."[30]

Justice Jackson called attention to the gap between this verdict and the broader context, suggesting through elegantly pointed phrasing that despite the legal tendency to keep this fact quiet, U.S. law often works hand in hand with the dominant majority to enforce its continued dominance: "There are limits to the extent to which a legislatively represented majority may conduct biological experiments at the expense of the dignity and personality and natural powers of a minority—*even those who have been guilty of what the majority define as crimes.* But this Act falls down before reaching this problem, which I mention only to avoid the implication *that such a question may not exist because not discussed.*"[31]

Race and Contemporary Biopolitics

While Skinner's sterilization test case may have seemed to put an end to eugenic interventions, which had by then become linked in the popular imagination to National Socialist race policies, in fact compulsory sterilization on eugenic principles continued, receiving serious public scrutiny only with the rise of the civil rights movement. While some states gradually decreased the rate of surgical sterilization (California, Kansas) other states actually increased the sterilization rate (North Carolina, Georgia, Virginia). According to Lombardo, nearly twenty-three thousand sterilization operations were "officially recorded in the United States between 1943 and 1959. . . . Arguments continued to be made in medical journals and the popular press that sterilization would prevent the transmission of 'hereditary diseases' and the birth of 'mentally deficient children'" (Lombardo 2008, 241–242).

Even when the impact of the Eugenics Record Office at Cold Spring Harbor, New York, seemed to have waned, and its 12.25 linear feet of Trait Card Boxes and 32.0 linear feet of Trait Files documenting traits as diverse as hair color, skin color, temperament, diseases, and deformities had been replaced by genetic and genealogical records held at the Charles Fremont Dight Institute for the Promotion of Human Genetics at the University of Minnesota and later at the Center for Human Genetics, the use of race as a category subject to management and intervention continued.[32] As legal scholar Dorothy Roberts has argued, the eugenics-based notion that science could be harnessed to improve individuals and populations has not been repudiated but merely shifted its zone of operations from the social management of reproduction and the enforcement of strict immigration policies to the more specialized context of biomedicine (Roberts 2006).[33] With the rise of genomics and a new interest in so-called genetic medicine, race management has been molecularized, rendered seemingly natural, and thus gained widespread acceptance. Lovalerie King has pointed out that Leon Higginbotham's two questions drew attention to the negative role played by racial distinctions in the U.S. legal system: "Did the law merely perpetuate old biases and prejudices? Or had it been an instrument first in establishing and only later in attacking injustices based on color?" (King 2007, 163n13). Now, responding to the medicalization of race, Dorothy Roberts argues that we must apply the "strict scrutiny test in equal protection law." We should reverse Higginbotham's inquiry and ask "the fundamental question, when does race consciousness in public policy legitimately further the state's interest in racial equality?" (Roberts 2006, 530).

Obama Bucks

The double-edged sword of race consciousness, which can lead both to Dorothy Roberts's argument for "strict scrutiny in equal protection law" and to Leon

Higginbotham's challenges to race-based legal bias, brings us back, finally, to those disturbing Obama Bucks. Created by a Democratic staffer and then disseminated by a Republican one, the image was ostensibly created to deplore negative racial stereotypes, but it became explosive precisely because it relied on just such a negative stereotype: the black chicken thief.

Not only did the image create a political tempest but it also made waves in the corporate world. Drawing attention to the racialized meaning of property is bad for business, it seems. Both Kentucky Fried Chicken and its Louisville owner, Yum Brands, as well as Kraft Brands, the conglomerate owner of Kool-Aid, issued strongly worded denunciations of their products' appearance (in the form of the bucket of Kentucky Fried Chicken and the pitcher of red Kool-Aid) in the Obama Bucks image. While the corporations' responses attest to their views that the Obama Bucks image was "very troubling and potentially legally actionable," in fact the bucket of chicken and the Kool-Aid are highly ambiguous images, just like the dancing donkey and the slice of watermelon (Levister 2008). While the Kentucky Fried Chicken is certainly most likely to be interpreted as a derogatory allusion to the food preferences of African Americans, it can also call to mind the inadequate access to nutritious food possessed by low-income people in urban areas, or it can even recall pressure put by members of the black community, such as Rev. Al Sharpton, on the industrial chicken producers to stop their inhumane treatment of animals. As *New York Times* reporter Melanie Warner explained, PETA (People for the Ethical Treatment of Animals) "was eager to enlist Mr. Sharpton because KFC has many stores in African-American communities and in late 2003 KFC executives told investors they were making an increased effort to market to African-Americans" (Warner 2005). And while the pitcher of Kool-Aid that smiles over the dancing donkey's left shoulder can be seen as invoking both the Jim Jones cult's mass suicide as well as the stereotype that African Americans have a special attachment to red Kool-Aid, it also speaks to the economic pressures felt by African American mothers, who must feed their children Kool-Aid rather than the more expensive milk, fruit juice, or soda.[34] Finally, in its very name the drink alludes to the special ability to stay cool that distinguishes "No Drama Obama" in the eyes of many pundits.

The implications of any image are plural, indeterminate, prismatic, of course. Among the innumerable images of black men with chickens that circulated during the run up to the 2008 election, most of them are mechanistic repetitions of familiar jokes or folk tales, such as the old question, "Why did the chicken cross the road?" the comparison of presidential candidates to Chicken Little (Chicken-licken of old) who called out "the sky is falling," and the old saying, "The chickens have come home to roost." But while we focus on the meaning of the stereotype something else escapes our notice: the fact that there is a long-lived, pejorative association between the chicken and the black man. And if we don't notice this fact, we are unable to explore its significance. As Ludwik Fleck so trenchantly observed in 1935, "We can never sever our links with the past, complete with all its errors. It survives in accepted concepts, in the presentation of problems, in the syllabus of formal education, in everyday life, as well as in language and institutions. That which has occurred in the past is a greater cause of insecurity—rather it only *becomes* a cause of insecurity—when our ties with it remain unconscious and unknown" (Fleck 1979, 30).

Examining one of our greatest national institutions, the inauguration ceremony, *Washington Post* reporter John Kelly reminded us of the etymological origins of the word "inauguration," and asked us to remember for a moment the mingled dread and fear with which the ancient Romans read the entrails of those sacred chickens to find what the future might bring. To my way of thinking, we should also contemplate the profane chickens linking African American men to racist stereotypes that have been with us since the era of slavery. Like the "Chicken Snatcher" toy and the race movies using variations of the chicken thief plot, the Obama Bucks image provides a troubling augury for the inauguration day of our forty-fourth president: it testifies to the complex node of racial identity and legal principle that has formed the racialized understanding of both property and theft embedded in the U.S. Constitution. A year after his inauguration, President Obama began the struggle to pass a comprehensive reform of the nation's health care system. That this struggle, too, would be marked by the continuing impact of the nation's racial past was augured only too clearly by one posting to a blog headed "Senate Passes Historic Health Care Changes": "Our 'president' would be all too happy to sign into law a plan that steals from those who have succeeded to give to those who haven't because there are a lot more slackers who haven't, and the successful people wouldn't have voted for a chicken thief like Obama anyway."[35]

Conclusion

Zen of the Hen

"In the beginner's mind there are many possibilities, in the expert's there are few."

— Suzuki Roshi, *Zen Mind, Beginner's Mind*

This book ends, as it began, with sitting—or to be exact, with Zen meditation. One morning in July 2005, the idea for a book appeared in my mind, chapters and all, while I was meditating. I was keeping chickens at the time; just a week before I had added to my flock the two Cobb 700 pullets whose short lives I chronicled in the introduction. Bred for meatier breasts and easier deboning, they weren't up to their hybrid breed standard, and had been scheduled for culling—that is until they were given to me by my friend at the Poultry Education Research Center at Penn State. The day I picked them up, I wrote this entry in the journal I keep next to my *zafu*, or meditation cushion: "Poor things are still babies—handle easily, and have naked, unfeathered stomachs. They were reared to be killed for meat at six weeks old. Awful."

Zen practitioners talk about "beginner's mind": the lack of preconceptions, eager openness, and anticipation with which a student approaches a new subject. Well, when I first planned this book, I was thinking with "expert's mind," and I planned it to be rather different from the book you hold in your hand. I felt pretty sure what I wanted to say in this book that came to me so suddenly. I imagined it as a history of poultry science in which I would show how the thriving academic discipline had fallen victim to the rise of corporate in-house science as the foundation for a critique of industrial poultry production.

During the five years of researching, writing, and living this book, although I have continue to mourn the decline of poultry science as a field and to protest the damaging global impact of industrial poultry farming, I have gradually sloughed off some of the preconceptions, settled opinions, and assumed expertise of that earlier researcher. I have encountered ambiguities, complications and challenges as I have explored chicken culture, and in response I have returned to something closer to what Buddhists call beginner's mind. "Beginner's mind is Zen practice in action. It is the mind that is innocent of preconceptions and expectations, judgments and prejudices. Beginner's mind is just present to explore and observe and see 'things as-it-is.' . . . Without approaching things with a fixed point of view or a prior judgment, just asking 'what is it?'" (Hartman 2001). That daily encounter with the question "what is it?" is essential to a journal, of course, so I turn now to a journal entry to demonstrate what it has meant to learn the Zen of the hen.

July, 2009: It's one thing to write about Nancy Luce's methods for doctoring her hens and quite another to find myself out in the chicken yard, with the old bare-butt Buff Wyandotte on my lap, trying to figure out why she's got poop all over what used to be a shiny bald pate of a butt. Turns out she had a prolapsed cloaca: the inner puffy tissue was bulging out about an inch. So I remembered Nancy Luce: "you have some wit and a slender finger" and I pushed the cloaca back in. It took a couple of tries, because it seemed to pop right out again; what did Nancy do about that? I wondered. Since then, I'm not sure my hen's cloaca has stayed entirely in its proper place, but she does seem to have a cleaner butt. And I have a less arrogant perspective on old Nancy Luce: she's not quite the naïve rural character she must have seemed to me before. Seems pretty brave and straightforward to fix a chicken that way. Or perhaps I'm seeing myself as much closer to Nancy Luce. I wonder who's older at this moment: Nancy as she was then, or me now, less than a year from my sixtieth birthday? There she was, all by herself, making a living out of her chickens and her poems, and feeling alone, mocked, irregular. And here I am, two centuries later, profiting from feminism and the "marriage privilege" to do what I want with only minimal mockery: my endowed professorship and married status keeping me from the ridicule of a spinster "chicken lady." Or so I like to believe.

And then there's Betty MacDonald, whose hatred of her hundred chicks I've just written about as a symptom of industrial farming, an emotion caused (I've argued) by the increase in scale and the male dominance of the new kind of farming she and her husband are trying out, in their postwar Vashon Island chicken farm. While all of that seems true to me, there is also another kind of truth that the last several weeks have really taught me: the impingement, frustration, and near-frenzy that comes over you when you have a lot of noisy chicks under your care.

My daughter and my husband built a lovely chicken tractor for me—was it only two and a half weeks ago?—as a solution to the problem of raising day-old hatchery chicks. It was a several-stage process. We had begun by keeping them in our garage, in the converted dog run with chicken-wire reinforced sides that we'd made for them before they arrived in the mail. That really worked remarkably well: we put down a blue tarp, covered it with wood shavings and paper towels for the first several days (to keep the newly hatched chicks from ingesting any wood chips in their random pecking) and then encircled it with the corrugated cardboard draft shield we'd been sent by Murray McMurray, the hatchery from which they came. Across the top we strung wire from which hung two heat lamps that we could raise or lower depending on their need, and all of it was capable of being closed at night when we lowered the garage door, or opened to sun and wind when we raised it.

After a couple of weeks it became clear that the chickens needed to get out into the open air. They were hopping about like popcorn, even flying over each other, and we decided to move the whole apparatus onto the lawn. We were planning to modify Joel Salatin's technique of pastured poultry, which would mean keeping the chickens on our non-chemical lawn (or "pasture") in our repurposed dog run, moving them daily, and enjoying the lawn fertilization, delicious eggs, and low feed costs that would effortlessly follow [Salatin 1993; see also Lee and Foreman, 2006]. So, we crated up all of the chickens one night and with the help of our neighbors moved the huge unwieldy structure over the split-rail fence with the fringe of deer-netted lilies, onto the lawn. And popped the chickens inside, to their delight. Flies! Clover! A view! We'd put the dog igloo inside (now dubbed—with apologies to Johannes Paul and his friends at Omlet—the eggloo) and covered half of the top with a tarp to give them some shelter. Of course the tarp flapped, which scared them, and the eggloo seemed to have little appeal except as something to peck (were they actually eating the plastic doorsill?) and to stand on. But it was workable.

We had the ideal set-up, we thought: shelter, grass, fresh air. But that was until we tried to move the massive structure. Two of us could do it, little by little, by edging each galvanized aluminum corner slightly to one side, and then running around to the other corner and doing the same. But then we were slipping in the copious amounts of chicken poop these birds produced. And we were also having to go inside to move the eggloo manually (hoping that the birds wouldn't slip out as we did so). And each time we also had to pick up all the feeders and waterers and refill them (cleaning them first to get the poop off because the birds just loved to sit on top of them, with gravity taking its toll). Yet left to my own devices there was no way I alone could move the chickens every day to give them fresh grass, and give our grass the chance to recover from their nitrogen onslaught. Finally it became clear that there had to be a better way to raise chickens.

So, on to phase two: the design and construction of a poultry tractor made of much lighter PVC pipe and chicken wire. My daughter (the designer) spent a whole day building the beast, as I tied off wires to hold the chicken wire to the PVC piping, supplied ice water, and baked a "thank you" berry and rhubarb pie. Then, late in the afternoon, we finally moved the chickens into their new home. Once again they hopped around happily, ate grass, and clustered at the wrong end, in the sun, rather than under the nice tarp we had thoughtfully arranged over the top of one end, or in the eggloo we had also, laboriously, moved inside the tractor.

I'll pass over the effort to make sure they were secure for the night, as well as the anxiety with which I awoke the following morning and ran out to see if they were still alive. What happened next was the problem: three or four days of moving the chicken tractor, light as it supposedly was, every day. That involved reaching in on hands and knees (hands and knees in chicken poop) to grab the feeders and waterers so they could be rinsed with a bleach solution and refilled, and then one person lifting the tractor quickly while the other with difficulty pushed the eggloo to one side so that it would shuffle to the new footprint. Neither act was easy on the back. Twice, a chicken decided to slip out to freedom while the tractor was being lifted. Luckily they are such symbionts that the chicken just hung around the outside gazing at its fellows and peeping, giving me time to grab it and shove it back into the tractor. I'm uncomfortable with the tone of contempt that seems to be slipping in here, but I guess it's bred from the exasperation of that complicated lifting, grabbing, and moving, while slipping all the time in chicken poop, with flies buzzing around.

It wasn't just the physical labor, either: it was the emotional work! I'd sit on the deck, trying to clear my mind to write, or to read Aldrovandi on chickens, and find that I was staring at the birds across the lawn. Did they have enough food? Were they too hot? Why weren't they going under the tarp shelter? Did they have enough water? Periodically our whippet would get the same obsession, and would run at them, or run around them, stirring up terror and forcing them all to run peeping into a little knot in one corner.

Finally, we decided to split the chickens up and put half of them in the chicken coop with the other hens and the other half in the poultry tractor which we would resituate in the meadow beyond our chicken yard. So, one more time: catch the chickens and plop them into the crate to move them. (This one-at-a-time task had now become pretty routine and two-at-a-time; as I tipped them into the crate I talked to them the whole time soothingly which seemed to help, though I'm not sure whether it was helping them or me.) Then we carried the poultry tractor on our shoulders through the front gate, down the driveway to the gap between the apple trees, back to the right into the meadow, and then up meadow until we slowly set it down, parallel to the apple tree next to the big hen house, but outside the chicken yard. Replaced the feeders, the waterers, and the tarp, and put the chickens back in. Once again, happy chickens; long grass, bugs, sun and wind.

But then came the first night out in the meadow. I'd read in Joel Salatin's book that chicken wire can't keep a raccoon out: the dexterous little varmint will pull a chicken apart by just reaching through the wire. But exhausted after all that refitting and moving, I decided to deal with that problem in the morning. Come sunrise, I woke up early and went right out to find several birds wandering in the uncovered front of the tractor near the food and water, but the rest of the birds all clustered together under the tarp at the back. Clustered together, eating something . . . Oh.

It was a chicken. Actually, it was part of a chicken: the bright yellow foot, the feathery leg, and then what looked like something from a kitchen counter–pulled translucent chicken flesh of some sort, in a pile. And the chickens were pecking away at it. We'd had our raccoon

encounter: I soon confirmed it when I discovered the rest of the chicken—
a heap of feathers, flesh, and blood just outside the tractor's chicken-wire
wall.

So once again I got the crate, put all of the surviving chickens in it,
and moved them into the big hen house, where I gave them food and
water and shut them in. Now, a new set of worries and a new set of problems to solve with jerry-rigged solutions. If these large white pullets are in
the big hen house, where will the old hens lay their eggs? And yet if we
crowd them into the little hen house, which we'd created by walling off
part of our garden shed with chicken wire, where will the bantams (whose
house it had been) roost for the night?

I had read somewhere that chickens had a three-day memory, a theory I was able to confirm as I worked through these bedtime prob-
lems.Three times at dusk I gathered the chickens up from where they
huddled, in a clotted white mass on the darkened ground of the chicken
run, to place them in the little hen house. Three times I let them out in
the morning, only to have them wind up, at night, unable to find where
they were supposed to roost at dusk. But on the fourth evening when I
came out to close them in, they were already lined up in the big hen
house, shoulder-to-shoulder on their perches. They had clearly sorted by
size, and were now in their chosen home.

By the end of the episode of the poultry tractor, I was abashed to
find that I understood very well Betty MacDonald's exhaustion as she sat
in her cabin on Vashon Island, overwhelmed by the demands of her
needy, hungry, noisy chicks. Simply to have our birds in the hen house
rather than the poultry tractor was a major relief: no more poop on our
lawn, no more worries about predators.

That was July 2009. Rereading the journal entry in February 2010, I notice how
my certainties and judgments have softened. I seem to have learned the Zen of
the Hen. Her lesson, like Suzuki Roshi's, is *just sit*. Sit until you have moved
beyond preconceptions and expectations, judgments and prejudices. Sit until
you can "just [be] present to explore and observe and see 'things as-it-is.'"
Looked at "as it is," the world of chickens and the human beings who work and
play with them is harder to separate cleanly into heartlessly efficient production or virtuously tender stewardship. Instead, it mingles pleasure and fascination, trial and error, exhilaration and empathy, as well as irritation and even
exhaustion.

A snowstorm has brought the eastern seaboard to a welcome standstill. The sun is blazing on deep snow outside my window, and I have to jump up from my computer and run out shouting to scare away the large red-tailed hawk that landed in the big pine tree beside the hen house. In the last month despite our best efforts—a scarecrow with a Day-Glo hunting vest, much human patrolling, and frequent visits by our dog to the chicken yard—we've lost fifteen birds, birds I have cared for, have written about, and whose blue-, green-, and brown-shelled eggs I have gratefully eaten for breakfast. I'm determined to protect the rest of my chickens from the hawk, a frighteningly, beautifully effective predator. So now I sit at my computer with an eye on the sky, looking for those wide wings that augur so badly for my little flock. Another lesson that life "is what it is."

I began this book writing about the role of augury in our cultural imagination: the practice of sacrificing sacred chickens to learn the will of the gods or watching the behavior of birds to see what the future will bring. The term *augury* itself has been assimilated so deeply into our cultural unconscious we no longer remember that a bloody act of animal sacrifice lies behind the dignified human pomp of presidential inaugurations. Yet augury and sacrifice are both still very much a part of human and chicken culture, and I want to end by returning to them.

As I wrote this book, I tried hard to stay open to the contradictions and tensions in my role as chicken keeper. Contradictions are not comfortable, however, and I have corresponded and conversed with people for whom my meat-eating, in particular, caused a tension too great to tolerate. I regret in particular how it cut short a beginning friendship with an artist whose paintings of discarded chicks and crated chickens being trucked to the butcher would have offered a powerful example of Adam Smith's notion of fellow-feeling, grounded in the complex embodiment of humans and other animals.[1] At the same time, I have come to know other people I admire: young girls and boys who are learning the fine points of chicken exhibiting, small town dwellers raising a handful of hens in their back gardens, a professor of entomology with a thriving second career growing organic meat, and a cancer center employee from Seattle who is teaching poultry raising to working-class inner-city teenagers and their families.

Over the years I have particularly enjoyed conversations with my Mennonite butcher, Eli Reiff. He speaks with passionate specificity about how to butcher chickens humanely and offers persuasive analyses of the shortcomings of engineered "meat birds" and critiques of the short-sighted economies of scale governing the world of "Big Chicken" he has voluntarily left behind. Thinking of Eli Reiff's work and world, I recall Donna Haraway's challenge to feminism, in *When Species Meet*: "to figure out how to honor the entangled labor of humans and animals together in science and in many other domains, including animal husbandry right up to the table" (Haraway 2008, 80). In this partial ABC, as I have permitted myself the position of curious amateur, I have hoped to honor the complex relationships that comprise chicken culture.

NOTES

Introduction: Why Chickens?

1. Cobb 500 Sales Brochure, 2008, http://www.cobb-vantress.com/Products/ProductProfile/Cobb500_Sales_Brochure_2008.pdf.

2. Cobb 700 Sales Brochure, 2007, http://www.cobb-vantress.com/Products/ProductProfile/Cobb700_Sales_Brochure_2007.pdf.

3. Ibid.

4. It was Paul Farley who told me that White had been a devoted chicken keeper. Farley, an employee of the Fred Hutchinson Cancer Center, is also the poultry educator for Seattle Tilth, an organization that offers free classes in gardening and poultry raising in an old tan brick building on the highest hill in Wallingford, Washington, originally the Convent of the Good Shepherd, a "home for wayward girls." Paul Farley, personal interview, December 16, 2005, Seattle, Washington. When I interviewed Farley in 2005, not only did he introduce me to what might be called the demographics of chicken raising in the Pacific Northwest, describing how they raise chickens in city backyards in Seattle, and how 4-H has a different makeup on Bainbridge Island than it does in the rest of King County, but he also tipped me off that White contributed the introduction to Jones's volume. He even followed up in the mail with a photocopy of his own copy of the essay.

5. "Country life has many meanings. It is the elms, the may, the white horse, in the field beyond the window where I am writing. It is the men in the November evening, walking back from pruning, with their hands in the pockets of their khaki coats; and the women in headscarves, outside their cottages, waiting for the blue bus that will take them, inside school hours to work in the harvest. It is the tractor on the road, leaving its tracks of serrated pressed mud; the light in the small hours, in the pig-farm across the road, in the crisis of a litter; the slow brown van met at the difficult corner, with the crowded sheep jammed to its slatted sides; the heavy smell, on still evenings, of the silage ricks fed with molasses. It is also the sour land, on the thick boulder clay, not far up the road, that is selling for housing, for a speculative development, at twelve thousand pounds an acre" (Williams 1973, 3).

6. Clare Hinrichs, personal communication, Boalsburg, PA, 2009.

7. "About the Agrarian Studies Program," 2010, http://www.yale.edu/agrarianstudies/real/aboutAS.html.

8. This process of allowing chickens to be "adopted" has been since been discontinued by PERC because the high cost of feed requires that they use every bird they hatch, whether or not they meet the high standards of an exhibition judge.

9. An abecedarium is also a poetic form, twenty-six words in alphabetical order, according to one blog: http://www.mediatinker.com/blog/archives/007759.html.

10. Rather than being thought of as linear or causal, the phenomenon of dependent origination is best thought of as a chain with twelve links: ignorance, formations (everything born or created), consciousness, name and form (or mind/body), our six senses, contact (the meeting between the object of sense, the senses, and consciousness), feeling, craving, clinging, becoming, birth, aging and death. Christina Feldman, "Dependent Origination," Portion of a talk presented at the Barre Center for Buddhist Studies.

11. The association is not mine alone. English writer Martin Gurdon chose *Hen and the Art of Chicken Maintenance* to capture the paradoxical frustrations and pleasures of his years as a novice poultry farmer (Gurdon 2004).

12. Lott continues, "In addressing my study to a variety of fields and disciplines, I mean not only to properly portray a complex phenomenon but to help solidify the claims of cultural studies as a practice" (6). I could say the same, but in truth my impulse was personal as well as institutional. Because I have found cultural studies lacking adequate attention to agriculture in general and chicken culture in particular, I wanted to give myself the chance to follow my interests wherever this topic led them.

13. As the AP news wire lead put it, "Scott Brown staged an upset in last week's Massachusetts Senate special election, in part, by pledging to be the 41st GOP vote against President Barack Obama's health care overhaul." http://www.digtriad.com/news/the-buzz/story.aspx?storyid=136541&catid=57.

Chapter 1: Augury

1. "Augury." Encyclopædia Britannica Online. 2010. http://www.britannica.com/EBchecked/topic/42813/augury.

2. "In [the] conceptual frame of reference of an ultimately unbroken certainty of knowledge of the lifeworld, unawareness is predominantly conceived as *not-yet* knowledge or *no-longer* knowledge, that is to say, *potential* knowledge" (Beck 1999, 124).

3. Of course, the insights of Beck and Proctor both hark back to Ludwik Fleck's crucial delineation of the origins of ignorance as well as knowledge in the restrictive force of scientific thought communities (Fleck 1979, 30).

4. "Because of its particular epistemological positioning between knowledge and unawareness, literature is able to hold open a zone of exploration that other mediations (political, social, scientific, and economic) foreclose. Literature . . . offers an alternative to the expert discourse that, as [Ulrich] Beck and [Anthony] Giddens demonstrate, has become socially and epistemologically dominant" (Squier 2004, 22, 78–79).

5. British Lion Eggs Web site, http://www.lioneggs.co.uk/. My thanks to Richard Nash for the cell phone call that alerted me to the Lion Brand eggs.

6. Assured Food Standards Web site, http://www.redtractor.org.uk.

7. A press release from Assured Food Standards in 2006 reported: "In view of recent efforts to simplify and strengthen the Red Tractor assurance mark, AFS joined forces with British Chicken Marketing to back a new campaign for Great British Chicken. . . . We are confident that this latest development, which is supported by the whole industry, will be of significant value to the British chicken industry at this challeng-

ing time. . . . Our programme will highlight to consumers the importance of buying British and will reassure them about the quality and standard of our products." British Poultry Council Web site, http://www.poultry.uk.com/food/assurance.htm.

8. The curious mixture of science and whimsy continued after the appearance of high-pathogenic avian influenza in British poultry flocks. An item posted on the British Egg Information Web site in response to the appearance of H5N1 in British turkey flocks announced that "the British Egg Industry Council (BEIC) has welcomed Defra's decision to order free range birds inside within a limited area in Suffolk and South Norfolk, as both appropriate and timely to product flocks from bird flu." BEIC promises that careful monitoring of the bird housing will combine high biosecurity with rigorous attention to the problem of bird well-being: "Birds are housed every night as a matter of course and so have plenty of space to move around and all their food and water is inside. Many producers will hang items like cabbages and CDs, which effectively act as 'toys' to keep the birds occupied during their short stay indoors, which will ensure that the birds do not suffer possible stress." http://www.britegg.co.uk.

Chapter 2: Biology

1. My thanks to Johannes Paul and William Wyndham for their time, and the whole Omlet crew for their hospitality. Interview with Johannes Paul and William Wyndham at Omlet headquarters, Wardington, Oxfordshire, UK, November 7, 2007.

2. http://www.Omlet.co.uk.

3. "Omlet." Brochure distributed by Omlet Ltd., Tuthill Park, Wardington, Banbury, OX 17 1RR.

4. For the relevant Waddington papers, see the file PP/FGS/C18–19, CMAC, Wellcome Institute for the History of Medicine.

5. My thanks to Claudio Stern for a discussion that added to my knowledge of Waddington and of avian biology. See also Stern 2002.

6. "Carrel's New Miracle Points Way to Avert Old Age," *New York Times*, September 13, 1913, SM3.

7. Ibid.

8. My gratitude to Anne Buchanan and Ken Weiss for patiently answering my many questions ranging from avian embryology to contemporary genetics.

9. "Army Chicken Quota Bars Civilian Sales," *New York Times*, December 2, 1944, 1.

10. Harland Loy Schrader Collection, National Agricultural Library, MC 334 Box 1.

11. USDA Poster, courtesy of the National Agricultural Library. American Poultry Historical Society Personal Collection of O.August Hanke, Watt Publishing Company, Mount Morris, Illinois. Box 2 of 2.

12. "A Million Eggs for a Million Soldiers," *National Poultry Digest*. IBCA [International Baby Chick Association] News Release, July 1941, 400.

13. "The Chicken-of-Tomorrow Story," billed as "a factual report on the nationwide, three year breeding program" prepared by H. L. Shrader, senior extension poultryman, U.S. Department of Agriculture; Prof. A. E. Tomhave, University of Delaware poultry specialist; and Karl Seeger, University of Delaware poultry pathologist, offers a different figure, which may reflect the other costs of the entire contest: "A&P Food Stores representatives offered to put up $10,000 in cash awards and to underwrite expenses of

operations." Personal files of John E. Wedlick, American Poultry Historical Society, Box 1 of 1, National Agricultural Library, Beltsville, MD.

14. See also "1951 National Chicken-of-Tomorrow Contest Results." Agricultural Experiment Station, University of Arkansas College of Agriculture, Fayetteville, Arkansas. December, 1951. Report Series 30. Personal Collection of Robert M. Smith, American Poultry Historical Society, Box 2. National Agricultural Library, Beltsville, MD.

15. *The Saturday Evening Post*, August 9, 1947, 116.

16. As W. Dewey Termohlen, director of the Poultry Branch of the Division of Production and Marketing of the USDA, put it, the goals of this new era were "economy of production, economy of marketing, quality of product, and merchandising." "Past History and Future Developments," *Poultry Science* 47 (1968), 30–31. Typescript in American Poultry Historical Society Personal Collection of O. August Hanke, Box 1 of 2, National Agricultural Library, Beltsville, MD. And the ideal chicken would be characterized by "fast and economical growth . . . high quality meat at a low price" (Hanke 1950, 82; see also Schlosser 2001).

17. "1951 National Chicken-of-Tomorrow Contest Results," 4.

18. Quote is from I. M. Lerner, letter to Melvin W. Buster, August 2, 1945. Personal collection of Robert M. Smith, Box 2, National Agricultural Library, Beltsville, MD.

19. Ibid.

20. Melvin Buster to I. M. Lerner, August 1945. Personal collection of Robert M. Smith, Box 2, National Agricultural Library, Beltsville, MD. "Directions for Preparing a Wax Model of a Dressed Chicken" recommended the use of bleached beeswax, and the suggestion that "those who are not in a position to undertake a casting job . . . have the work done commercially. In most of the larger cities, business firms that do this work can be located through the Classified Section of the telephone directory, under the heading of 'Sculptors.'" Personal collection of Robert M. Smith, Box 2, National Agricultural Library, Beltsville, MD.

21. See also I. M. Lerner, letter to Melvin W. Buster, August 2, 1945. Personal collection of Robert M. Smith, Box 2, National Agricultural Library, Beltsville, MD.

22. Buster assured them that "[we] do not want to make any unreasonable requests or urge you to undertake any difficult task the results of which could not be utilized in connection with your projects and teaching work." Melvin Buster to R. M. Smith, September 26, 1945. Personal collection of Robert M. Smith, Box 2, National Agricultural Library, Beltsville, MD.

23. Ironically, this bird was bred by Charles Vantress, a UC Berkeley–educated poultry breeder. Gordon Sawyer's discussion of the Chicken-of-Tomorrow contest winners offers a fascinating window into the contest's effects on poultry breeding, showing how hybrid birds (or crosses) came to dominate poultry breeding because of the increased public demand for such birds that the contest produced (Gordon Sawyer 1971, 118–122). Vantress founded one of the premier hybrid chicken breeding corporations today, the global poultry corporation Cobb-Vantress. http://www.cobb-vantress.com/AboutUs/.

24. A pamphlet in the archives of the National Agricultural Library presenting the "Scoring Procedure for the 1948 National Chicken-of-Tomorrow Contest" itemizes the winning traits. The highest score—20 points—was reserved for two criteria, "Body well proportioned" and "Breast: broad, long, full-meated, well-rounded," while a "plump, full-meated" thigh and a "full-meated" drumstick, and a "wide, long, flat, well-

fleshed" back were each worth ten points. The score card includes a caveat: "In the final judging in the national contest, the judges may give consideration to the percent of edible breast and leg meat in making the final awards."

25. American Poultry Historical Society Collection of Glossy Prints from Salisbury Laboratories in Charles City, IA. Box 1 of 2. National Agricultural Library, Beltsville, MD.

26. American Poultry Historical Society Personal collection of Robert M. Smith, Box 2, National Agricultural Library, Beltsville, MD.

27. Harland Loy Schrader Collection, National Agricultural Library, Beltsville, MD. Box 1.

28. Ibid.

29. American Poultry Historical Society Personal Files of John E. Wedlick, Box 1 of 1.

30. Ibid. Another example in the poultry history archives is the "1951 National Chicken-of-Tomorrow-Contest Results," whose cover photograph shows a smiling woman in a cashmere sweater holding a chick, superimposed on a photograph of two large poultry houses. The caption reads, "The front cover pictures the two brooder houses at the University of Arkansas Farm, built by the poultry industry especially to handle the 1951 National Chicken-of-Tomorrow Contest. The inset shows Miss Joan Walters, of Rogers, Ark, queen of the 1951 carnival." "1951 National Chicken-of-Tomorrow Contest Results." Agricultural Experiment Station, University of Arkansas College of Agriculture, Fayetteville, Arkansas. December 1951. Report Series 30. Personal Collection of Robert M. Smith, American Poultry Historical Society, Box 2. National Agricultural Library, Beltsville, MD.

31. *The poultry man national newspaper of the industry*, Friday, February 4, 1949, 1. Personal files of John E. Wedlick, American Poultry Historical Society Box 1 of 1, National Agricultural Library, Beltsville, MD.

32. The film follows the fortunes of an Arizona family whose mother is forced to take in boarders when the get-rich-quick schemes of her shiftless husband meet with disaster. New York *Herald Tribune*, January 31, 1949, np. Clipping held in the Personal Files of John E. Wedlick, Box 1 of 1, American Poultry Historical Society. National Agricultural Library, Beltsville, MD. See also the review by Bosley Crowther, who described the film as "a diet of good, substantial cooking in the Hollywood sentimental style" but concludes that "it is just a bit disappointing that Mr. Seaton . . . should have let this 'Chicken Every Sunday' stew along in a much warmed pot" (Crowther 1949).

33. N.a., n.t., *New York Herald Tribune*, January 31, 1949; see also *The poultry man national newspaper of the industry*, Friday, February 4, 1949, 1. Personal files of John E. Wedlick, American Poultry Historical Society Box 1 of 1, National Agricultural Library, Beltsville, MD.

34. *New York Herald Tribune* January 31, 1949, n.p. Emphasis mine.

Chapter 3: Culture

1. Stephen Green-Armytage, "Extraordinary Chickens 2007 Wall Calendar."

2. I choose this as my horizon of exploration because, like Marilyn Strathern, I see it as having a characteristic set of economic, scientific, social, cultural, and agricultural structures for understanding and tending chickens. Of course, to limit this to Euro-American culture is a drastic limitation, especially when one is discussing a bird whose origins lie in Indonesia and that holds immense spiritual and social significance for so much of the non-Euro-American world. But because I am focusing

primarily in this chapter on British and American trends in poultry breeding and exhibiting as well as the works of an American photographer, I have made this reluctant limitation. Jean Pagliuso does not stand in for artists from other parts of the world, but her approach to the chicken portrait is certainly suggestive of trends one might follow up in other parts of the world.

3. Sarah Franklin has written with wit and perception about this phenomenon, in *Dolly Mixtures* (2007).

4. See also "Obituary: Mr. William Bernhard Tegetmeier." *Ibis* 55, no. 1: 136–138. doi:10. 1111/j.1474–919X.1913.tb06546.x.

5. The publication of these volumes continued until *The American Standard of Perfection 2001* (Burgettstown, PA: American Poultry Association, 2001), now out of print.

6. Darwin singles out the Cochin, with its deeply furrowed skull bones and "peculiar voice," as the only breed that might have been "descended from some unknown species, distinct from *G. bankiva*" (Darwin 1868, 268).

7. That particular belief may be on its way back as I write this, since food price increases, linked to the increasing cost of corn-based feeds and petroleum-based transportation, seem to have reintroduced the idea of backyard poultry as an economic venture. Yet in the intervening years, Prebleman's outburst to his wife—"Honey, they're a steal! We could stick a bunch of these in the barn, feed 'em table scraps—it'd cut our grocery bills in half!"—would have seemed no more sensible than the notion of raising chinchillas (Perelman 1952, 131).

8. Jean Pagliuso, interview with the author, New York City, New York, December 18, 2006. All citations are from this interview. My gratitude to Jean Pagliuso for her generosity in granting me several long interviews in her studio.

9. Jean Pagliuso Photography Web site, http://www.pagliuso.com/people/index.html. As Pagliuso explained, the first photographs she took after leaving commercial photography were "teeth. Really large scale teeth that [she] bought in the medina in Marrakesh. Pulled human teeth." Pagliuso interview. From fashion to teeth to fragile remains: the sequence suggests some of the competitive atmosphere in the fashion world, recently satirized in the film *The Devil Wore Prada*.

10. Unfortunately, she discovered, her father's chickens had all been given away to "this feed store that dad used to buy all his feed from, and they just gave them away to kids." Still, she was determined to do some chicken photographs; the birds had come to seem "kind of an interesting object."

11. Press Release, Marlborough Graphics, available at http://www.marlboroughgallery. com. Indeed, by bragging that the show featured "more than twenty types of chickens," the gallery revealed its naïveté: the correct term would have been "breed." If the emphasis on technique and aesthetic lineage represented an attempt to justify the appearance in this major Manhattan gallery of a show concerned with so "unlikely" a preoccupation as the breeding of "show chickens," a similar tension was evident when the show moved to the Drawing Room in East Hampton. When that show was reviewed by *Art Forum* in its online diary, high fashion redeemed low agriculture: "'Fowl as couture,' New Museum director Lisa Phillips observed, and indeed each chicken looked somehow coiffed and styled by the likes of Chanel, Balenciaga, and Dior. Leave it to an artist to transform a barnyard animal into a thing of *beauté*" (Yablonsky 2006).

12. According to Baker, this term was coined by Gustav Fredrich Hartlaub "as the title for a 1925 exhibit at the Mannheim art gallery consisting of art that rejected the frag-

mentation of impressionism and Expressionism in favor of, as Hartlaub put it, 'loyalty to a positively tangible reality'" (Baker 1996, 75–76).

13. For a stunning series of Sander's photographs with accompanying commentary, see The Getty Museum Web site: http://www.getty.edu/art/gettyguide/artMakerDetails? maker=1786.

14. As the first and last chapters of this study explore at greater length, the augurs of ancient Rome foretold the future by observing the behavior of chickens. The Etruscan soothsayers known as haruspices played a less political role. They, too, prophesied the future, but their study was the entrails of sacrificial animals (Smith 1875, 586–587). See also Brittanica Online, "Haruspices," http://www.britannica.com/ EBchecked/topic/256270/Haruspices; and Benjamin 1972, 25. Benjamin's essay appeared first in *The Literarische Welt* of September 18, September 25, and October 2, 1931.

15. To Alan Sekula, Sander's work is "methodologically implicated in the operations of social power and domination" (Jones 2000, 4), whereas to George Baker, Sander's work is "*necessarily* torn between the forces of narrativity and those of stasis," between repressive and liberating ideologies (Baker 1996, 72).

16. In chapter 8 we will see the ultimate biosocial implications of this flexibility, as explored in the multimedia art of Koen Vanmechelen.

Chapter 4: Disability

1. When it was first performed at La MaMa in 1981, an earlier version of the play received little notice. After two additional performances in Italy in the 1980s, D'Ambrosi reworked the play completely, and he premiered the new version in Italy at the Teatro Patologico Festival, at the Nuovo Colosseo Theater, Rome, Italy (Slaff 2007).

2. I transcribed the play while I watched it, trying to capture the effect of the play in performance. All of the citations are as I heard them, although I have not been able to verify their accuracy (D'Ambrosi 2007). I was able to talk with Dambrosi after the performance, and he generously promised to send me the script of *Days of Antonio*, but sadly I never received it. Repeated attempts to contact him through Theater La MaMa or his agent also failed.

3. Dambrosi's historical accuracy slips here, for lobotomy was invented in 1935 by Walter Freeman and James Watts. For an illuminating analysis of the history and cultural meanings of lobotomy, see Jenell Johnson's dissertation, "Echoes of the Soul: A Rhetorical History of Lobotomy" (2008).

4. As Georges Canguilhem explains in his influential study, *The Normal and the Pathological:* "An anomaly can shade into disease, but does not in itself constitute one" (1991, 140.

5. As bioethicist Mark Kuczewski explains in a brave if only partially successful attempt to reconcile the discourses of disability studies and bioethics, "If this environment is unyielding and unaccommodating to the point that important life activities are denied to the person with an impairment, that impairment becomes a handicap" (Kuczewski 2001, 40).

6. Compare to Hughes's earlier works, "Medicine and the Aesthetic Invalidation of Disabled People" (2000), and "The Constitution of Impairment: Modernity and the Aesthetic of Oppression" (1999).

7. Paul K. Longmore, posting to the Disability Studies in the Humanities listserv, July 1, 2003. DS-HUM@LISTSERV.UMD.EDU.

8. Jessa Chupik recommends Joseph W. Schneider's "Afterword" to *Incurably Romantic* by Bernard F. Stehle (Philadelphia: Temple University Press, 1985) for a discussion of "academic work that ignores individuals with intellectual disabilities and describes 'the class system of disability.'" Jessa Chupik, posting to the Disability Studies in the Humanities listserv, January 31, 2003. DS-HUM@LISTSERV.UMD>EDU January 31, 2003.

9. Richards describes Geoffroy as a French transcendental anatomist, whereas Melinda Cooper describes him as a natural historian, and Jane M. Oppenheimer locates him in natural philosophy as well as natural history, demonstrating the rhetorical force of disciplinary location in crafting histories of science.

10. "Pour ne pas étendre indéfiniment le cercle de mes recherche, je me suis borne a une seule espèce. Les œufs de poule, que l'on peut se procurer en nombre aussi considérable que l'on veut, se prêtent avec la plus grande facilité à l'étude de l'évolution normale et de l'évolution tératologique. L'incubation artificielle, que l'on peut employer aujourd'hui d'une manière scientifique, permet de faire des expériences absolument impossibles dans l'incubation naturelle.

    On verra plus tard comment les résultats que j'ai obtenus, par l'étude d'une seule espèce, s'appliquent en réalité à tout l'embranchement des animaux vertébrés."

11. "They tried to produce cyclopia but said they could not see very well the parts on which they were producing the lesions. The heads of the resulting embryos were highly atypical, if present at all; but there were sometimes two hearts present. . . . Their general conclusions . . . were that the inversion of head and of heart are not consequences of one another, as Dareste had once postulated, but separate effects of parallel causes, related to the unequal growth of the two sides of the embryo" (Oppenheimer 1968, 156).

12. "Recessive inheritance was first revealed in alkaptonuria (1902), an enzyme deficiency that leads to cartilage degeneration, and albinism (1902). Dominant inheritance was discovered in brachydactyly (short fingers, 1905), congenital cataracts (1906), and Huntington's chorea (1913). And sex-linked inheritance was discovered in Duchenne muscular dystrophy (1913), red-green color blindness (1914), and hemophilia (1916)." "Hereditary Disorders," Image Archive on the American Eugenics Movement, http://www.eugenicsarchive.org/eugenics/list_topics.pl.

13. Jan Witkowski, "Traits Studied by Eugenicists," 7. Dolan DNA Learning Center, Eugenics Archive. http://www.eugenicsarchive.org/html/eugenics/essay_4_fs.html; and M. Cooper 2004, 20. See also Canguilhem 1991.

14. As he formulated them, those updated principles are
    1. Identical defects can be induced by administering different agents.
    2. Particular embryos react to an adverse condition in dissimilar ways.
    3. The dissimilarity is caused by unequal combinations of inherited gifts and extrinsic influences.
    4. Type of defect depends upon the strength and time of action of adverse impulse.
    5. The smaller the defect, the later it becomes apparent (Jelenk 2005, 296).

15. See the photograph, "Chicken coops and first hybrid corn (behind) at Carnegie Station for Experimental Evolution, Cold Spring Harbor, NYC," Harry H. Laughlin Papers, Truman State University, Lantern Slides, Brown Box, 2004. ID # 1021. Available at Dolan DNA Learning Center Web site, http://www.eugenicsarchive.org.

16. To Michael D. West, "the story is clearly a product of the year of 1918." West reports that Anderson wrote his friend Trigant Burrow on New Year's Eve 1918 to announce the impending appearance of the short story: "The poor little magazine called *The Little Review* will be publishing in a month or two a tale of mine called The Triumph of the Egg." Anderson later withdrew the story from *The Little Review* and it appeared in *The Dial* in March 1920 (West 1968, 676–677).

17. For compelling discussions of the role of freak shows in the construction of disability, see Thomson 1996. Although she does not directly define the term "freak," Thomson offers a rich description of how disabled characters are made into freaks in literary texts by being "stripped of normalizing contexts and engulfed by a single stigmatic trait" in her monograph, *Extraordinary Bodies* (1997). For biographies and photographs of some of the disabled actors in *Freaks*, see "Freaks of Nature," The Missing Link Web site, http://www.missinglinkclassichorror.co.uk/freaksnatttxt.htm. And for photographs of the Coney Island freak show members who were visited by field workers from the Eugenics Record Office under Charles Davenport's leadership, see the Dolan DNA Learning Center Web site, http://www.eugenicsarchive.org/eugenics/view_image.pl?id=1014.

18. The Missing Link Web site, http://www.missinglinkclassichorror.co.uk/freaks.htm.

19. "For example, two recruits were Frances O'Connor, an armless woman known professionally as 'The Living Venus de Milo,' and Minnie Woolsey, whose body is believed to have been affected by a rare disorder that causes premature aging and who was billed as 'Koo Koo, the Bird Girl from Mars'" (Norden and Cahill 1998, 88). For more on the cast of *Freaks*, see Adams 1991.

20. According to the Missing Link Web site, the horrific resonances of this scene extended to the costumes as well; "the chicken-suit used for Baclanova's final scene was originally made for Browning's 1928 Lon Chaney vehicle, *West of Zanzibar*." "The Missing Link Reviews Tod Browning's *Freaks*" http://www.missinglinkclassichorror.co.uk/freaks.htm.

21. See Squier 2004 for a discussion of C. P. Snow's pseudonymously published novel about pharmaceutical interventions in the human life course, *New Lives for Old* (London: Camelot, 1933). Snow's novel was published one year after *Freaks* was released in the United Kingdom, where it "was a disaster at the box office and a heavy blow to the studio's reputation. Indeed, the film drew many highly negative reviews, prompted civic groups across the country to renew calls for movie censorship, and was banned outright in the United Kingdom" (Norden and Cahill 1998, 88–89).

22. I discuss regenerative medicine at greater length in *Liminal Lives* (2004). See also Melinda Cooper, *Life as Surplus* (2008); and Nikolas Rose, *The Politics of Life Itself* (2006).

23. Responding to *Lucy*, a play offered by the E.S.T. and Alfred P. Sloan Foundation's Science and Technology in 2007 in New York City, Jessica Ruvinsky argued, "It IS possible that autism could be the next stage in evolution. As long as there is variation, it's heritable, and it leads to differential reproductive success—that is to say, if slightly-autistic geeks get more play—then natural selection may increase the frequency of autism in the future" (Ruvinsky 2007). For more on enhancement technology see Elliott 2004.

24. A reclamation of the earlier tradition of teratology, with its affirmation of the possibilities of the anomalous, also suggests ways to move beyond the historical trajectory of disability to date, as traced by Rosemarie Garland Thomson, within which "the

wondrous monsters of antiquity . . . became the fascinating freaks of the disabled people of the later twentieth century. The extraordinary body moved from portent to pathology" (Thomson 1997, 58).

25. *Yolk* (2007) Head Pictures, produced by Damon Escott, writer/director Stephen Lance. Funded by the Australian Film Commission under the "Short Film" fund. Lena is played by Audrey O'Connor. *Yolk* synopsis, Sprout Film Festival Web site, 2008. http://gosprout.org/film/info6.htm.

26. "Lena, a fifteen-year-old girl with Down Syndrome, is fascinated by sex and all the mystery that surrounds it, fertility, desire, and the mechanics of childbirth. Because of her inability to grasp the facts of sex, the desire for it envelopes Lena like a cloud of unknowing. Everywhere she looks, she sees sex. Her brother has a girlfriend who cannot keep her hands off him. The chicken egg she is caring for as part of a dimly understood school project has a vague, but ominous sex life of its own. Daniel, the boy next door has changed from innocent neighborhood playmate to the object of Lena's desire." *Yolk* synopsis, Sprout Film Festival Web site, 2008. http://gosprout.org/film/info6.htm.

27. I am paraphrasing Gregory Bateson, of course, who defines information as "differences that make a difference" (1979, 99).

28. Thanks to Kate Schaub for suggesting this satisfying reading when I presented this chapter at the Columbia University Medical School faculty seminar on narrative genetics.

Chapter 5: Epidemic

1. I am referring here to a way in which a kind of "folk wisdom" girds Hoggart's and Williams's work in, respectively, *The Uses of Literacy* (1998 [1957]) and the famous essay "Culture Is Ordinary" (1989 [1958]).

2. Department of Health and Human Services, "Key Facts about Avian Influenza (Bird Flu) and Avian Influenza A (H5N1) Virus," May 7, 2007. Center for Disease Control Web site. http://www.cdc.gov/flu/avian/gen-info/facts.htm.

3. "Nation's Leading Alarmists Excited about Bird Flu," *The Onion*, February 2, 2005. Issue 41:05. http://www.theonion.com/content/node/30868.

4. Federal News Service, Secretary of Defense Donald Rumsfeld Media Availability. News Transcript, May 22, 2002. United States Department of Defense, http://www.defense.gov/Transcripts/Transcript.aspx?TranscriptID=3457.

5. United States National Science Advisory Board for Biosecurity, meeting minutes, November 21, 2005. http://www.biosecurityboard.gov/meetings/NSABB%20November%202005%20meeting%20minutes%20-%20Final.pdf. Accessed July 31, 2007.

6. U.S. Department of Health and Human Services, "ABC TV Movie: Fatal Contact: Bird Flu in America." http://www.pandemicflu.gov/news/birdfluinamerica.html.

7. Similar language can be found at the Web site of the 7th International Bird Flu Summit, http://www.new-fields.com/birdflu7/index.php.

8. Similar language can be found at the Web site of the 7th International Bird Flu Summit, http://www.new-fields.com/birdflu7/index.php?p=papers.

9. "Animal husbandry," *Dictionary.com Unabridged* (v. 1.1). Random House, Inc. http://dictionary.reference.com/browse/chicken.

10. At the time this book was going to press, a self-styled "femivore" movement that advocated a back-to-the-land lifestyle of free-range chicken raising and organic gardening had achieved some carefully cultivated media presence. As Peggy Orenstein (2010) of the *New York Times* observed, this return to chicken raising as the responsibility of a good wife seemed less a liberation into better food policies than a return to a bad old past. "It was an unnervingly familiar litany: if a woman is not careful, it seems, chicken wire can coop her up as surely as any gilded cage."

11. The poultry industry workforce shifted from its predominantly African American composition in the 1970s to Mexican and Central American immigrants during the 1980s and by the turn of the next century relied so heavily on an immigrant workforce that the U.S. Justice Department filed a series of suits against the Tyson Corporation for smuggling illegal immigrants from Mexico to supply fifteen poultry-processing plants in the U.S. South (Striffler 2005, 98).

12. "How Cobb Has Become World Leader," http://www.cobb-vantress.com/AboutUs/CobbHistory.aspx.

13. Less fitting perhaps, given the arduous physical work that such industrial poultry practices require, is the fact that "Siloam" is a biblical term meaning a place where healers minister unto the afflicted.

14. "CobbSasso 150: The natural choice," http://www.cobb-vantress.com/Products/Cobb-Sasso150.aspx. This bird is an excellent example of nature being denatured and renatured, a process explored in illuminating detail by Sarah Franklin, Celia Lury, and Jackie Stacey (2000, 73–74).

15. "Cobb 700 on Target for Low-Cost, High Meat Yield." *Cobb Focus* 2 (2002): 1.

16. Of course, the industrial model still remains one among several, alongside the family farm and the small farm. Yet industry, government agencies such as the Peace Corps, and even nongovernmental organizations such as the Heifer Project are exporting "the U.S. model of factory farming of chickens" to other parts of the world (see Sachs 1996). To recap, in addition to concentration on market breeds such as the Cobb 700, this model relies on contract growers, who provide the property, the buildings, and the risk and who are provided by the industry with the chicks, feed, and management protocols to raise them until they are sold back to the industry at a price that fluctuates with the international market.

17. However, the story does include among the beneficiaries of the accomplishment "comparative genomicists desiring to accurately identify the functional elements of the human genome; and genome-sequence producers" (Schmutz and Grimwood 2004, 679).

18. By 2009 the tide had turned against chickens and eggs as front line in the fight against influenza, though the target now was swine flu, or H1N1. The press bemoaned the lag times built into live-egg vaccination production and pharmaceutical companies promised to produce vaccines faster by alternative means. "Wonder Why H1N1 Shots Are Late? Blame Eggs," AP, updated October 21, 2009. http://www.msnbc.msn.com/id/33420826.

19. "Research Program," http://www.cobb-vantress.com/AboutUs/ResearchProgram.aspx.

20. The GAO report suggests that the authors found industry cooperation lacking when they were compiling their report. For example, the report cites the problem of inadequate record keeping by OSHA and alleged "underreporting in the meat and poultry industry" (GAO 2005, 43–44).

21. "University of Iowa Health Prediction Markets: Avian Influenza," Influenza Prediction Market 2007, http://fluprediction.uiowa.edu/fluhome/Market_AvianInfluenza.html. Recently this project acquired a new name, the Iowa Electronic Health Markets, and its Web site acquired a new look. See http://iehm.uiowa.edu/iehm/index.html.

22. "The National Animal Identification System (NAIS)" USDA Animal and Plant Health Inspection Service, July 17, 2007. http://www.aphis.usda.gov/newsroom/content/2007/07/usaionais.shtml.

23. "USDA Announces New Framework for Animal Disease Traceability," Release No. 0053.10, February 5, 2010, http://www.usda.gov/wps/portal/usda/usdahome?contentidonly=true&contentid=2010/02/0053.xml.

24. Patricia Dunn, DVM, interview with author, April 5, 2006.

25. "The possible link to Smithfield has not been reported in the U.S. Press," Tom Philpott reported. "Searches of Google News and the websites of the New York Times, Washington Post, and Wall Street Journal all came up empty . . . I'll be curious to see whether the U.S. media explores the link with Smithfield's Mexico operation" (Brainard 2009).

26. David Kirby responded, in the comments to the Brainard blog post: "I have made it clear that we do not know where this virus originated, as you indicated. Even in my current piece online, I state that it is 'entirely possible' the virus did not come from a Mexican CAFO, but it is not a 'wild theory' either, as Reuters labeled it in a story yesterday. I still think there is plenty of evidence to continue investigating Mexico and swine CAFOs (on all continents) because this virus had to have mutated somewhere into its current form—and residents of La Gloria began complaining of strange flu symptoms in February, if not sooner. Only one four-year-old boy's sample was tested, and it came back positive for the new H1N1 strain. It could be a coincidence that he was the only person in town to get the disease, and stranger things have happened, but it remains suspect. Meanwhile, testing of pigs may reveal nothing, because the virus may have moved out of the herd by now, or carrier pigs may have died or been killed. I believe that workers should also be tested for low-level, asymptomatic infections, which is more likely to occur among people who are exposed to zoonotic pathogens on a daily basis and develop some immunity, animal scientists told me. Which leads me to my final point. You note that bloggers 'did not call experts to test their theories.' I can't speak for others, but I spent hours interviewing officials from the CDC, USDA and the US Trade Office, as well as Dr. Ellen Silbergeld, professor of environmental health sciences at Johns Hopkins Bloomberg School of Public Health, and a leading researcher of pathogen evolution in CAFOs; Dr. Liz Wagstrom, director of veterinary science at the National Pork Board; Dr. Gregory Gray, a University of Iowa professor of international epidemiology and expert in zoonotic infections; Bob Martin, former executive director of the Pew Commission on Industrial Farm Animal Practices and currently a Senior Officer at the Pew Environmental Group; plus some U.S. hog growers and other people involved in agricultural science. All of them said the La Gloria connection was a plausible explanation, and some thought it was likely." http://www.cjr.org/the_observatory/swine_flu_and_cafos.php.

27. Schmidt was quoting Gregory Gray, director of the Center for Emerging Infectious Diseases at the University of Iowa College of Public Health.

28. "Current State of Vaccine Availability and the Next Steps in Production and Distribution Efforts," Statement of Ben Machielse, DRS., Executive Vice President, Operations, MedImmune, before the House Energy and Commerce Subcommittee on Oversight

and Investigations and Subcommittee on Health, November 18, 2009. http://energy-commerce.house.gov/Press_111/20091118/machielse_testimony.pdf.

29. The manual of standards for the production of pharmaceuticals and other medically active compounds is known as the European Pharmacopoeia. Established by the Council of Europe and published (online and in hard copy) in regularly revised versions, this manual stipulates that SPF chickens must be used for vaccine production. "Chicken Flocks Free from Specified Pathogens for the Production and Quality Control of Vaccines," *European Pharmacopoeia* 5.1, 5.2.2. 04/2005:50202, 2825. Available at http://www.edqm.eu/en/Online_Publications-581.html.

30. As this report defined them, SPF flocks are "isolated flocks with appropriate biosecurity controls. All birds are tested at least once for the range of pathogens listed in the European Pharmacopoeia either at point of lay (introduced SPF birds) or by 20 weeks of age (new generation birds within established flock)" (2). The report went on to explain, "The recent breakdown (September 2005) of one of only two Australian SPF flocks highlights the urgent need for end users to be able to access imported eggs during critical shortages" (5).

31. The "Contingency Import Policy" explains: "In reality, the diseases of quarantine concern are highly infectious and would spread rapidly. . . . For small flocks (i.e., less than 4,000 birds) monitoring the source flock's routine European Pharmacopoeia test results for 8 weeks post egg collection provides an equivalent confidence. . . . . However, to account for larger flocks . . . monitoring results for 12 or more weeks may be necessary to achieve an equivalent level of confidence" (Biosecurity Australia 2006, 11).

32. As the Web site of the Factory Farming Campaign of the Humane Society of the United States explains, "as male chicks cannot lay eggs and aren't the same breeds as those chickens raised for meat, they are of no value to the egg or broiler industries. Shortly after hatching, hundreds of millions of unwanted male chicks are killed by the commercial egg industry annually, usually by gassing, crushing, or suffocation." http://www.hsus.org/farm/multimedia/gallery/layers/male_chicks.html.

Chapter 6: Fellow-Feeling

1. "Tyson to Sell Chicken Free of Antibiotics," *New York Times*, June 20, 2007. http://www.nytimes.com/2007/06/20/business/20tyson.html.

2. "The Natural Choice," Cobb Sasso 150 sales brochure, http://www.cobb-vantress.com/Products/ProductProfile/Cobb_Sasso150_Sales_Brochure.pdf.

3. Lawrence Birken's investigation of the notion of psychic economy in Freud's work arises from the psychoanalytic project of reconciling neoclassical political economy with the realm of seemingly irrational human behaviors: "If Freud appeared to depose the principle of the utility-maximizing individual by offering numerous exceptions to that principle, he actually showed that these exceptions were only apparent on the surface level of consciousness. . . . Freud claimed to have expanded the sphere of social science so that final determination [of the routes of energy in the individual brain] no longer rested upon the conditions of production (as the Marxists claimed) but upon the material conditions of reproduction" (Birken 1999, 324). In short, Birken argues, Freud envisioned psychoanalysis as a "kind of unified field theory in which the objects of the political economy originally based on satisfying non-sexual needs were revealed to be a mere subset of the larger 'libidinal economy' in which energy

could be invested in oneself, one's relatives, one's wife, or a shiny new bicycle down the street. The only qualitative distinction which thus remained was that between a life–affirming consumption and the extinction of consumption itself in death" (327). Bill McKibben's *Deep Economy* (2007) seems to take off from this point, exploring how consumption has now begun to produce death, and thus we need to redirect both aspects of *homo oeconomicus*, the economic and the deeper psychosexual, from death to life.

4. Here Sugden departs from Castiglia (2002) in viewing the turn to depth and affective internalization expressed in Smith's *Theory of Moral Sentiments* not simply (and inevitably) as a mode of surveillance and control designed to perpetuate existing inequities. Rather, he argues that the recognition of affect has the potential to provide an alternative model for conceptualizing, assessing, and assigning value in society. In this alternative model, value is determined not by positivist quantification of instrumental benefits but by affective assessments of "subjective but non-instrumental" goods (Sugden 2002, 81). Think, perhaps, of the nation of Bhutan's decision to index "gross national happiness" (see Schell 2002).

Chapter 7: Gender

1. M. G. Kains explained in *Profitable Poultry Production*: "Formerly hens were regarded as a necessary nuisance, tolerated mainly because they lay the foundation of custards, cakes and other dainties, the enjoyment of which offsets somewhat the losses of grain and garden truck. . . . It is little wonder that poultry raising has had difficulty in shaking off the disrepute in which it was formerly held. The whole trouble has been in the mental attitude of the farmer. This has subjected the fowls to systematized neglect. Hens relegated to the stables, wagon sheds, fences or trees for roosting places; to the mow or the manger for nests; to the barnyard and field for feed, cannot do well. . . . This condition of affairs is happily being replaced by better management, because better management pays" (Kain 1910, 6). Kain is equally dismissive about the economic value of chicken keeping: "The cost and value of the eggs consumed at home is rarely considered by the general farmer. Hens are kept because the housewife must have eggs for making certain dishes as well as for boiling, poaching, frying, etc. If they were not kept the farmer would either have to do without or purchase eggs. As the former does not suit his palate nor the latter his pocketbook, he tolerates a few hens which care for themselves more or less, and which pick up a considerable amount of forage that would otherwise go to waste. If they supply the family's needs he is content to consider the yield in eggs and chickens as offsetting his losses of grain which he has to feed the flock" (16).

2. I have adapted the phrase, inserting the word "agriculture" in place of Waldby's word, "medicine."

3. United Food and Commercial Workers Women's History Timeline, http://www.ufcw. org/womens_history_month/timeline/index.cfm.

4. Both the 1928 and 1957 versions of *The Little Red Hen* are unpaginated.

5. The poultry industry reliance on artificial incubation resulted in the preferential breeding of "non-broody" hens, so that even rare-breed hatcheries such as Murray McMurray specify hens that will make good mothers as an exception worth noting, as in this description of the Buff Orpington: "They also make excellent setters and mothers." Murray McMurray Hatchery, http://www.mcmurrayhatchery.com/product/buff_orpingtons.html.

6. "Hens Mated Artificially in Jersey Experiment," New York Times, October 5, 1937. http://select.nytimes.com/gst/abstract.html?res=F50B12FD34541B728DDDAC0894D841 5B878FF1D3.

7. Ibid.

Chapter 8: Hybridity

1. As critics Barbara Simons and Wouter Keirse put it, "Why the Bresse, precisely? And why the Malines [the Mechelse Koekoek]? By choosing such distinctive varieties of chicken, Koen Vanmechelen immediately nailed his colors to the mast. He views the two birds as the (all but commonplace) personification of the typical Frenchman and the typical Fleming" (Simons and Keirse 2003, 25).

2. The apt metaphor comes from Johan Buelens, a political scientist at the Free University of Brussels (Amies 2007).

3. "Where science would eliminate mutations of this kind," Simons and Keirse observe, "the 'Cosmopolitan Chicken Project' treats them with the dignity that Vanmechelen believes they deserve. After all, he feels, it is not by denying its limitations but by embracing them that the human spirit is able. . . . to break out of the cage that holds it prisoner" (Simons and Keirse 2003, 37).

4. Venice Projects, Ltd., "Koen Vanmechelen," http://www.veniceprojects.com/?Lang=_ 1&id_pagina=136&id_pagina_2=42&id_scheda=49&pag=1.

5. This map appears to be either a Mollweide equal-area map or one of its adaptations. For more on the Mollweide and other projections, see Komzak and Eisenstadt, 2001.

6. Reflecting Vanmechelen's linguistic hybridity as well as his densely private set of interpretations, the list under "Conclusion" reads: "—the primal chicken stands high up the ladder of the evolution and defied many centuries;—Monogamy might be there [sic] strength;—Universal diseases remain limited;—Inbreeding within a certain group is only occasionally;—No inherited errors by natural selection;—Domestication leads to vulgar behavior;—Diseases an inbreeding are spread by polygamy;—The creation of different races in different countries is a call for stepping over to maintain as a species;—The creation of a super bastard is actually a reconstruction of the creation by will;—The triangle is opposite direction" (Vanmechelen 2008, 4).

7. This projection was developed by German astronomer and mathematician Karl Mollweide (Vanmechelen 2008, 4).

8. Given Vanmechelen's preoccupation with Flemish linguistic and cultural identity, Mercator might well have the status of an iconic figure.

9. The resolution read in part: "WHEREAS, world maps have a powerful and lasting effect on peoples' impressions of the shapes and sizes of lands and seas, their arrangement, and the nature of the coordinate system, and WHEREAS, frequently seeing a greatly distorted map tends to make it 'look right,' THEREFORE, we strongly urge book and map publishers, the media and government agencies to cease using rectangular world maps for general purposes or artistic displays. Such maps promote serious, erroneous conceptions by severely distorting large sections of the world, by showing the round Earth as having straight edges and sharp corners, by representing most distances and direct routes incorrectly, and by portraying the circular coordinate system as a squared grid. The most widely displayed rectangular world map is the Mercator (in fact a navigational diagram devised for nautical charts), but other rectangular world maps proposed as replacements for the Mercator also display a greatly distorted image of the spherical Earth" (Rosenberg 2009).

10. This phrase is drawn from Vujakovic's survey of the "broad range of contemporary discussions of the political significance of maps" in which he mentions in particular "Yi-Fu Tuan's (1974) discussion of centrism in the maps of various cultures" and "Harley's (1989) 'rule of ethnocentricity'—which he suggested is 'a kind of "subliminal geometry," [adding] geopolitical force and meaning to representation'" (Vujakovic 2003, 62).

11. Varadarajan is deputy editor of the New Delhi newspaper, *The Hindu*, as well as the author of the blog "Reality, One Bite at a Time: India, Asia and the World."

12. The most famous usage is that of Diogenes, who when asked his origin, famously said, "I am a citizen of the world" (Kleingeld and Brown 2009).

13. I thank Irina Aristarkhova for sharing with me her illuminating rethinking of the whole notion of Kantian hospitality in terms of an occluded matrixial materiality. Her exploration of the role of the excluded woman in the concept of a cosmopolitanism grounded in hospitality parallels the role of the excluded chicken—the actual *Gallus gallus bankiva*—in Vanmechelen's construction of a cosmopolitan chicken.

14. As Fine defines "the new cosmopolitanism," "it is a way of thinking that declares its opposition to all forms of ethnic nationalism and religious fundamentalism as well as to the economic imperatives of corporate capitalism. It . . . aims to reconstruct political life on the basis of an enlightened vision of peaceful relations between nation states, human rights shared by all world citizens, and a global legal order buttressed by a global civil society" (Fine 2003, 452).

15. As they describe this new methodology of inquiry, it is a "trans-disciplinary one, which includes geography, anthropology, ethnology, international relations, international law, political philosophy and political theory, and now sociology and social theory" (Beck and Sznaider 2006, 2).

16. "A New Map of the Earth," *New Internationalist: The People, the Ideas, the Action in the Fight for World Development*, no. 124 (June 1983). http://www.newint.org/issue124/contents.htm.

17. As Peter Vujakovic points out, Peters argued for his own projection on the basis that it offered an objective, scientifically accurate representation of the globe. Focusing on the errors in the Mercator projection and its rounded-grid method, Peters claimed in contrast that the rectangular grid method he used represented scientific mode of cartography in its fidelity to the shape of the actual earth. However, as he points out, "From a political perspective on cartography . . . Peters' map is no less a rhetorical device, grounded in the attitudes and values of its author, than the maps and the associated value systems that he criticized" (Vujakovic 2003, 62).

18. He further argued against the idea that sterility was a result of natural selection, which according to creationists protected distinct species by rendering hybrids sterile (Darwin 1859, 231).

19. The term "hybrid vigor" comes down to us even now to describe the pound mutt, thought to be so much healthier than the overbred purebred: see chapter 9 in this volume for a discussion of the role of race and hybridity in the public understanding of President Barack Obama. Yet in his first White House press conference, President Obama told reporters that the "mutt" category was posing a problem for them as they were searching for a dog for their daughters: "our preference is to get a shelter dog, but obviously, a lot of the shelter dogs are mutts like me" (Kornreich 2008). Obama's comment implies that a mutt would most likely not be hypoallergenic.

20. Darwin continued: "and this is the conclusion of Mr. Blyth, and of others who have studied this bird in India." See also Ridley 2006, 37–47.

21. Darwin's cousin, Francis Galton, an enthusiast of the study of heredity, had recently published the curious study *Hereditary Genius*, "a collective autobiography of the masculine Victorian elite," according to Darwin's biographer Janet Browne. Not surprisingly, Galton found *Variations* inspiring; unfortunately, the tests he carried out to confirm that study's theory of "pangenesis" ended up disproving it, and the article documenting those tests published in *Nature* was a further blow to the book's standing (Browne 2002, 290, 291–292).

22. "Selection may be followed either methodically and intentionally, or unconsciously and unintentionally. Man may select and preserve each successive variation, with the distinct intention of improving and altering a breed, in accordance with a preconceived idea; and by thus adding up variations, often so slight as to be imperceptible by an uneducated eye, he has effected wonderful changes and improvements. It can, also, be clearly shown that man, without any intention or thought of improving the breed, by preserving in each successive generation the individuals which he prizes most, and by destroying the worthless individuals, slowly, though surely, introduces great changes" (Darwin 1868, 1:4).

23. *Variation*, a study so exhaustive in its documentation that the research for it taxed Darwin's mental and physical health, was his attempt to fill out the omissions that so troubled him in *Origin*. As he wrote, "No one can feel more sensible than I do of the necessity of hereafter publishing in detail all of the facts, with references, on which my conclusions have been grounded" (Darwin 1859, 11).

24. He reports that he has consulted "an eminent authority, Mr. Sclater, on this subject, and he thinks that I have not expressed myself too strongly" (Darwin 1868, 1:245).

25. While Desmond and Moore's book overstates its case in arguing that Darwin's early encounter with an African bird taxidermist and his travels on the *Beagle* solidified a lifelong commitment to antiracist scientific research, they are correct in their assertion that nineteenth-century race science made productive use of the longstanding analogy between the chicken and the human being.

26. These terms were invented by George R. Gliddon, one of the "anti-black, pro-slavery, separate-homeland pluralists" (Desmond and Moore 2009, 288–289).

27. "Darwin Was Wrong about Wild Origin of the Chicken, New Research Shows." *Science Daily*, March 3, 2008. http://www.sciencedaily.com/releases/2008/02/080229102059.htm.

28. "We cannot exclude the possibility that yellow skin was introgressed to the red junglefowl by hybridization with grey junglefowl prior to domestication, but it is much more plausible that introgression was facilitated by human activities. The red and grey junglefowls are full species as demonstrated by the fact that hybridization does not occur in the wild and when attempted in captivity, only a cross between grey cocks and red hens produced mostly sterile offspring. Hybridization between grey junglefowl and domesticated fowl, however, have been reported in the vicinity of villages within the area of contact between the two wild species, suggesting that the introgression of yellow skin into domestic birds took place after chickens were initially domesticated" (Eriksson et al. 2008, 4).

29. Eriksson and colleagues' research clarified, through phylogenetic analysis, the fact that the domestic chicken was of hybrid origin: "though the white skin allele originates from the red junglefowl, the yellow skin allele originates from a different species, most likely the grey junglefowl" (Eriksson et al. 2008, 2).

Chapter 9: Inauguration

1. My thanks to Lovalerie King, whose book *Race, Theft, and Ethics: Property Matters in African American Literature* (2007) inspired this chapter.

2. Psyche Williams-Forson includes some stunning examples of the genre from her own personal collection in a chapter with the pointed title, "Who Dat Say Chicken in Dis Crowd: Black Men, Visual Imagery, and the Ideology of Fear" (Williams-Forson 2006, 38–79).

3. As William H. Wiggins Jr. points out, the caption recalls Louis Jordan's 1940s record, "Ain't nobody here but us chickens!" and reappears in "countless chicken-stealing jokes" (Wiggins 1988, 249).

4. My thanks to Guna Nadarajan for bringing this toy to my attention.

5. This image appears in Williams-Forson 2006, 54.

6. "Zip Coon," *Uncle Tom's Cabin & American Culture*, The Institute for Advanced Technology in the Humanities, available at http://www.iath.virginia.edu/utc.

7. Lott is citing Douglass's 1849 article on Gavitt's Original Ethiopian Serenaders, a black minstrel company (Douglass 1849.) Reprinted in Foner 1950–1975, 1:141–142. Arguably, Barack Obama's own confrontation with the complexities of race in his memoir, *Dreams from My Father: A Story of Race and Inheritance* (2004), is shadowed by the same history of stereotype, bigotry, and resistance. Eric Lott uses the term *race movies* in his essay, "Love and Theft: The Racial Unconscious of Blackface Minstrelsy" (1992), and Gerald R. Butters Jr. also uses the term in his study *Black Manhood on the Silent Screen* (2002, 6). I use the terms *race films* or *race movies* interchangeably in this text. After the first appearance, when I place the terms in quotations to indicate their origins, I will refrain from using quotation marks.

8. Jane Gaines argues that "the significance of race movies would be the way in which they could be counterhegemonic without symmetrically 'countering' white culture on every point; for their oppositionality, if it could be called that, was in the circumvention in the way they produced images that didn't go *through* white culture" (Gaines 2001, 13).

9. Curiously, this was not the Biograph Company's first venture into chicken films. In the *Biograph Bulletin* for August 29, 1903, Sprocket Film No. 2401, "Strictly Fresh Eggs," was summarized: "A country house-wife is seen in her kitchen breaking eggs into a cake dish. As each egg drops it turns into a small chicken much to the woman's amazement and annoyance" (Niver 1971, 88). This surreal film image seems a remarkable homage to the important part played by eggs in the history of embryology as well as anticipating Sherwood Anderson's erstwhile chicken farmer now amazing his restaurant customers with egg tricks. Robert McKimson's mid-twentieth-century Looney Tunes cartoon character Foghorn J. Leghorn combines a chicken (or rather, rooster), the South, and Reconstruction-era politics in a stentorian romp that combines the racialized subtext with the seemingly inevitable turn to Kentucky Fried Chicken characteristic of other chicken images in twentieth-century United States. An entire book could be written analyzing the Foghorn Leghorn cartoons, and sadly this is not that book. See Keith Scott, *The Origin of Foghorn Leghorn* (2008) http://www.cartoonresearch.com/foghorn.html.

10. The other characters are, in toto, "A Southern White Farmer, His Boy, Four Neighboring White Farmers, Three Colored Farmhands, Four Colored Women, A Colored Clergyman, Two Pickaninnies" (Niver 1971, 140–143).

11. As film melodrama returned in the late 1970s to reconsider slavery from the perspective of the civil rights movement, Alex Haley's twelve-hour television miniseries *Roots* offered a further iteration of the chicken thief in the character of Chicken George. This mixed race descendent of the enslaved Kunta Kinte eventually earns his emancipation and by his "belated homecoming rescues his family from a race war that would have been suicidal" (L. Williams 2001, 249).

12. As Butters puts it, with great restraint, "As the [film] industry expanded exponentially and became more professionalized, African-Americans were deliberately left out of the equation, as in other industries that proved to be profitable" (Butters 2002, 27).

13. In its almost gleeful reallocation of property, this plot recalls the story of Luke, the escaped slave in Harriet Jacobs's *Incidents in the Life of a Slave Girl*. Luke has run away to the North after his master, Henry, has died. Like the slave in *Chased by Bloodhounds*, he has been given the old trousers of his erstwhile master, an act of penny-pinching pseudo-charity that Luke too has anticipated, and from which he as profited. As he tells Harriet Jacobs,

    "I'd been workin all my days fur dem cussed whites, an got no pay but kicks and cuffs. So I tought dis nigger had a right to money nuff to bring him to de Free States. Massa Henry he lib till ebery body vish him dead; an ven he did die, I knowed de debbil would nab him, an vouldn't vant him to bring his money 'long too. So I tuk some of his bills, and put 'em in de pocket of his ole trousers. An ven he was buried, dis nigger ask fur dem ole trousers, an dey gub 'em to me." With a low, chuckling laugh, he added, "You see I didn't steal it; dey gub it to me. I tell you, I had mighty hard time to keep de speculator from finding it; but he didn't git it."   (Brent 1987)

14. United States. U.S. Constitution, Amendment 14, *U.S. Constitution: Amendment 14*. The U.S. Constitution Online, http://www.usconstitution.net.

15. Also on the syllabus were the cases *State v. Mann*, the Fugitive Slave Act of 1850, and *Dred Scott v. Sanford.*

16. Links to President Obama's Constitutional Law course documents are provided in Kantor 2008.

17. Barack Obama, "Final Examination," *Constitutional Law III*—Fall 2001; scanned document, link found at Kantor 2008.

18. Ibid., 2.

19. Barack Obama, "Final Examination," *Constitutional Law III*—Autumn 2003; scanned document, link found at Kantor 2008.

20. See Weinbaum 2004 for a trenchant discussion of the ways these terms are deployed to produce and maintain racial and sexual inequality.

21. "Trait Files," *Eugenics Record Office Records, 1670–1964*, America Philosophical Society. Available at http://www.amphilsoc.org/library; and "Clinical and Field Studies of the 1921 Training Class," Eugenics Record Office, *Eugenical News* 6, no. 61, #1885, Eugenics Record Office, availabe at http://www.eugenicsarchive.org/eugenics.

22. Jan Witkowski, "Overview," Eugenics Record Office http://library.cshl.edu/archives/eugrec.html.

23. "Trait Files," *Eugenics Record Office Records, 1670–1964*, America Philosophical Society. Available at http://www.amphilsoc.org/mole/view?docId=ead/Mss.Ms.Coll.77-ead. xml#d4820643e3496103378944.

24. "Abstract." *Eugenics Record Office Records, 1670–1964,* American Philosophical Society. Available at http://www.amphilsoc.org/mole/view?docId=ead/Mss.Ms.Coll.77-ead. xml. The illustration "Eugenics" (broadside) is from the Harry H. Laughlin files, 1915–1938.

25. *Skinner v. State of Oklahoma Williamson,* 316 U.S. 535, 62 S.Ct.110, 86 L.Ed. 1655. *Open Jurist.* http://openjurist.org/316/us/535/skinner-v-state-of-oklahoma-williamson.

26. Ibid.

27. Ibid. The insertion here is mine. In Anglo-American property law, a bailee is a person who holds or acts as custodian of specific goods with which he or she has been entrusted subject to a contractual agreement. It is the responsibility of the bailee to return the article when the term of the contract is over. http://www.britannica.com/EBchecked/topic/49239/bailee.

28. "We have not the slightest basis for inferring that the line has any significance in eugenics nor that the inheritability of criminal traits follows the neat legal distinctions which the law has marked between these two offenses. . . . The equal protection clause would indeed be a formula of empty words if such conspicuously artificial lines could be drawn" (*Skinner v. State of Oklahoma*).

29. *Skinner v. State of Oklahoma Williamson,* 316 U.S. 535, 62 S.Ct.110, 86 L.Ed. 1655. *Open Jurist.* http://openjurist.org/316/us/535/skinner-v-state-of-oklahoma-williamson.

30. Ibid.

31. Ibid. My italics.

32. Eugenics Record Office Records. http://www.amphilsoc.org/mole/view?docId=ead/Mss.Ms.Coll.77-ead.xml.

33. Roberts also refers to Appiah 1990; and Lopez 1996. Barack Obama also draws on Appiah's work in his 1994 syllabus for "Current Issues in Racism and the Law." See Kantor 2008.

34. "When I drank Kool Aid as a kid, red (strawberry or cherry—never raspberry) was always preferred. Kool Aid is another marker of class. My parents couldn't afford fruit juice or bottles of soda for that matter for their eight offspring, so we drank Kool Aid as a substitute. I struggled as a single mother, but my kids could have fruit juice most of the time. I haven't bought a package of Kool Aid in at least 25 years." Lovalerie King, personal communication.

35. Gstanford on *Alienbabeltech.com,* February 4, 2009. http://alienbabeltech.com/.

Conclusion: Zen of the Hen

1. Sunaura Taylor Web site, http://www.sunaurataylor.org/portfolio/view/watercolors.

REFERENCES

Adams, Rachel. 1991. *Sideshow U.S.A.: Freaks and the American cultural imagination* Chicago: University of Chicago Press.

Agamben, Giorgio. 2002. *The open: Man and animal.* Stanford, CA: Stanford University Press.

Alpers, Paul. 1997. *What is pastoral?* Chicago: University of Chicago Press.

Amies, Nick. 2007. "Belgian political crisis and talk of linguistic divisions continue." *Deutsche Welle*, February 10. http://www.dw-world.de/dw/article/0,2144,2804788,00.html.

Anderson, Sherwood. 1920. "The triumph of the egg." *The Dial: A Semi-Monthly Journal of Literary Criticism, Discussion, and Information* (March): 4–13.

Anon. 1914. "The story of Chicken-licken." *The Waverly Juvenile Chapbooks*, no. 5, pp. 1–15. Illus. Philip Lyford (Boston: Seaver-Howland Press).

Appel, Toby A. 1987. *The Cuvier-Geoffroy debate: French biology in the decades before Darwin.* New York and Oxford: Oxford University Press.

Appiah, A. 1990. "Racism." In *Anatomy of racism*, ed. David Theo Goldberg, 3–16. Minneapolis: University of Minnesota Press.

Baker, George. 1996. "Photography between narrativity and stasis: August Sander, degeneration, and the decay of the portrait." *October* 76 (Spring): 72–113.

Barbato, G. F. 1991. "Genetic relationships between selection for growth and reproductive effectiveness" *Poultry Science* 78:444.

Bateson, Gregory. 1979. *Mind and nature: A necessary unity.* New York: Dutton.

Baynton, Douglas. 2001. "Disability and the justification of inequality in American history." In *The new disability history*, ed. Paul K. Longmore and Lori Umansky, 33–57. New York: New York University Press.

Beck, Ulrich. 1992. *Risk society: Towards a new modernity.* London: Sage Publications.

———. 1999. *World risk society.* London: Polity Press.

———. 2000. *The brave new world of work.* Trans. Patrick Camiller. London: Polity Press.

Beck, Ulrich, and Natan Sznaider. 2006. "Unpacking cosmopolitanism for the social sciences: A research agenda." *British Journal of Sociology* 57, no. 1: 1–23.

Bellairs, Raymond. 1953. "Studies on the development of the foregut in the chick blastoderm. 1. The presumptive foregut area." *Journal of Embryology and Experimental Morphology* 1:115–124.

———. 1967. "Experimental twinning and multiple monsters in chick embryos." In *Teratology: Proceedings of a symposium organized by the Italian Society of Experimental Teratology*, 162–173. Amsterdam: Excerpta Medica Foundation.

Benjamin, Walter. 1972. "A short history of photography." *Screen: The Journal of the Society for Education in Film and Television* 13, no. 1 (Spring): 5–26. (Essay orig. pub. 1931.)

Bernstein, David. 2008. "The Volokh Conspiracy." May 3: "Obama as Constitutional Law Professor," http://volokh.com/archives/archive_2008_04_27–2008_05_03.shtml# 1209848068.

Berry, Wendell. 1996. *The unsettling of America: Culture and agriculture.* Berkeley, CA: Sierra Club Books.

Bertelli, Aldo, and Luigi Donati, eds. 1969. *Teratology: Proceedings of a symposium organized by the Italian Society of Experimental Teratology.* Amsterdam: Excerpta Medica Foundation.

Bérubé, Michael. n.d. "Disability studies. Exploring the full range of human life." Rock Ethics Institute, Penn State University. http://rockethics.psu.edu/documents/disability.pdf.

Bhabha, Homi. 1994. "Signs taken for wonders." In *The location of culture.* New York: Routledge.

Biosecurity Australia. 2006. "Contingency import policy for specific pathogen free (SPF) chicken eggs." January 2. Canberra, Australia: Department of Agriculture, Fisheries and Forestry.

Birken, Lawrence. 1999. "Freud's 'economic hypothesis': From homo oeconomicus to homo sexualis." *American Imago* 56, no. 4 (1999): 311–330.

Bourdieu, Pierre. 1984. *Distinction: A social critique of the judgment of taste.* Cambridge, MA: Harvard University Press.

Boutang, Pierre-Andre, and Michel Parmat, dir. 1988. *L'abécédaire de Gilles Deleuze.* With interviewer Claire Parnet. *Art and Popular Culture Encyclopedia,* http://www.artand-popularculture.com/L%27Ab%C3%A9c%C3%A9daire_de_Gilles_Deleuze.

Bowen, Dana. 2007. "Old MacDonald now has a book contract." *New York Times,* June 20, D6. http://www.nytimes.com/2007/06/20/dining/20farm.html.

Boyd, William. 2001. "Making meat: Science, technology, and American poultry production." *Technology and Culture* 42, no. 4 (October): 631–664.

Boyer, Michael K. 1895. *Money in hens.* Syracuse, NY: C. C. DePuy.

Bradbury, Ray. 1948. "The inspired chicken motel." In *I sing the body electric.* New York: Alfred A. Knopf.

Brainard, Curtis. 2009. "Swine flu and CAFOs?" The Observatory. *Columbia Journalism Review,* April 29. http://www.cjr.org/the_observatory/swine_flu_and_cafos.php.

Brandth, Berit. 2002. "On the relationship between feminism and farm women." *Agriculture and Human Values* 19, no. 2: 107–117.

Brent, Linda [Harriet Jacobs]. 1987. *Incidents in the life of a slave girl: Written by herself,* ed., Jean Dagan Yellin. Cambridge: Harvard University Press (Orig. pub. 1861.)

Britt, Robert Roy. 2009. "This is not the Porky Pig plague." *LiveScience Health,* April 29. http://www.livescience.com/health/090429-swine-flu-name.html.

Broglio, Ronald. 2008. *Technologies of the picturesque: British art, poetry, and instruments 1750–1830.* Lewisburg, PA: Bucknell University Press.

Browne, Janet. 2002. *Charles Darwin: The power of place.* Princeton, NJ, and Oxford: Princeton University Press.

Bryson, Michael A. 200. *Visions of the land: Science, literature, and the American environment from the era of exploration to the age of ecology.* Charlottesville: University of Virginia Press.

Bubnoff, Andreas von. 2005. "The 1918 flu virus is resurrected." *Nature* 437 (October 6): 794–795.

Buchanan, Anne. 2009a. "Darwinism without Darwin: The origin of species without natural selection?" *The Mermaid's Tale* [blog]. October 6. http://ecodevoevo.blogspot.com/2009/10/darwinism-without-darwin-origin-of.html.

————. 2009b. Personal communication, e-mail, October 13.

Bugos, Glenn. 1992. "Intellectual property protection in the American chicken-breeding industry," *Business History Review* 66, no. 1 (Spring): 127–268.

Burgdorf, Robert L., Jr., and Marcia Pearce Burgdorf. 1977. "The wicked witch is almost dead: *Buck v. Bell* and the sterilization of handicapped persons." *Temple Law Quarterly* 50:995–1033.

Burnham, George Pickering. 1855. *A short history of the hen fever: A humorous record.* Boston: James French and Company.

Burrows, Montrose T. 1910. "The cultivation of tissues of the chick-embryo outside of the body." *Journal of the American Medical Association* 55, no. 24: 2057–2058.

Butters, Gerald R., Jr. 2002. *Black manhood on the silent screen.* Lawrence: University Press of Kansas.

Cameron, Jenny, and J. K. Gibson-Graham. 2003. "Feminizing the economy: Metaphors, strategies, politics." *Gender, Place & Culture: A Journal of Feminist Geography* 10, no. 2: 145–157.

Canguilhem, Georges. 1991. *The normal and the pathological.* New York: Zone Books. (Orig. pub. 1966.)

Canti, R. G. 1932. "Cinematograph Work." CMAC: SA/SRL 23/H.2 (Strangeways Research Laboratory Scientific 1932 Cinematograph Work).

Carpenter, Cliff D. 1962. Address to the Sydney, Australia, Rotary Club. August 14.

Carrel, Alexis, and Montrose T. Burrows. 1910a. "Cultivation of adult tissues and organs outside of the body." *Journal of the American Medical Association* 55, no. 16: 1379–1381.

————. 1910b. "Cultivation of sarcoma outside of the body: A second note." *Journal of the American Medical Association* 55, no. 18: 1554.

Casey, Jane Galligany. 2009. *A new heartland: Women, modernity, and the agrarian ideal in America.* Oxford: Oxford University Press.

Castiglia, Christopher. 2002. "Abolition's racial interiors and the making of white civic depth." *American Literary History* 14:32–59.

Chatterley, Cedric N., and Alicia J. Rouverol, with Stephen A. Lord. 2000. "*I was content and was not content": The story of Linda Lord and the closing of Penobscot Poultry.* Carbondale and Edwardsville: Southern Illinois University Press.

Cheah, Pheng. 2006. *Inhuman conditions: On cosmopolitanism and human rights.* Cambridge, MA: Harvard University Press.

Clark, Nigel. 2002. "The demon-seed: Bioinvasion as the unsettling of environmental cosmopolitanism." *Theory Culture Society* 19:101–125.

Clarke, Adele. 1998. *Disciplining reproduction: Modernity, American life sciences, and "the problems of sex."* Berkeley: University of California Press.

Cloke, Paul J., and Terry Marsden. 2006. *Handbook of rural studies.* London: Sage Press.

Clough, Ben. 2006. "Poor Nancy Luce." *The New Colophon* II (Part Seven): 253–265.

Conlogue, William. 2001. *Working the garden: American writers and the industrialization of agriculture.* Chapel Hill: University of North Carolina Press.

Cooper, Dani. 2008. "Scientists in spat over ancient chicken DNA." *ABC Science Online,* July 30. http://www.abc.net.au/news/stories/2008/07/30/2318696.htm?site=science&topic=latest.

Cooper, Melinda. 2004. "Regenerative medicine: Stem cells and the science of monstrosity." *Medical Humanities* 30, no. 1: 12–22. http://mh.bmj.com; doi:10.1136/jmh.2003.000137.

————. 2008. *Life as surplus: Biotechnology and capitalism in the neoliberal era.* Seattle: University of Washington Press.

Courson, Paul, and Steve Turnham. 2003. "Amid furor, Pentagon kills terrorism futures market," *CNN*, July 30. http://www.cnn.com/2003/ALLPOLITICS/07/29/terror.market/index.html.

Crenshaw, Kimberlé, 2005. "Intersectionality and identity politics: Learning from violence against women of color." In *Feminist theory: A reader*. ed. Wendy K. Kolmar and Frances Bartkowski, 533–542. 2nd ed. Boston: McGraw-Hill.

Crowther, Bosley. 1949. "Celeste Holm and Dan Dailey star in 'Chicken Every Sunday,' New bill at the Roxy." *New York Times*, January 19.

D'Ambrosi, Dario. 2007. *Days of Antonio (I Giorni di Antonio)*. Premiere performance, La MaMa Experimental Theatre Club, New York City, December 20.

Dareste, M. Camille. 1891. *Recherches sur la production artificielly des monstruosités ou essais de tératogénie expérimentale*. Paris: C. Reinwald & Cie, Librares-Editeurs.

Darwin, Charles. 1859. *The origin of species by means of natural selection or the preservation of favored races in the struggle for life*. New York: The Modern Library.

———. 1868. *The variation of animals and plants under domestication*. 2 vols. London: John Murray; reprinted New York: D. Appleton and Co., 1900.

———. 1871. *The descent of man and selection in relation to sex*. London: John Murray.

Davidson, Paul. 2009. "Pork producers buffeted by unfounded link to flu." *USA Today* November 11. http://www.usatoday.com/money/industries/food/2009–11–11-hogfarms 11_ST_N.htm.

Davis, Lennard J., and David B. Morris. 2007. "Biocultures manifesto." *New Literary History* 38:411–418.

Davis, Mike. 2005. *The monster at our door: The global threat of avian flu*. New York: New Press.

DeNoon, Daniel J. 2009. "Swine flu vaccine at least six months away." *WebMD*, April 30. http://www.webmd.com/cold-and-flu/news/20090430/swine-flu-vaccine-at-least-6-months-away.

Desmond, Adrian, and James Moore. 2009. *Darwin's sacred cause*. New York: Houghton Mifflin.

Donaldson, Elizabeth. 2005. "The psychiatric gaze: Deviance and disability in film." *Atenea* 35, no. 1 (June): 31–48. http://ece.uprm.edu/artssciences/atenea/Atenea-XXV-1.pdf.

Donaldson, Susan V. 2006. "Introduction: The Southern Agrarians and their culture wars." In Twelve Southerners, *I'll take my stand: The South and the Agrarian tradition*, 75th Anniversary Edition, ix–xl. Baton Rouge: Louisiana State University Press.

Douglass, Frederick. 1849. "Gafitt's original Ethiopian serenaders." *North Star*, June 29.

Drevenstedt, J. H. 1898. *The American standard of perfection, as adopted by the American Poultry Association*. Burgettstown, PA: American Poultry Association.

During, Simon. 1993. *The cultural studies reader*. 2nd ed. London: Routledge.

Ebert, Roger, 2006. "The real dirt on Farmer John" (review), *Chicago Sun-Times*, January 20. http://rogerebert.suntimes.com/apps/pbcs.dll/article?AID=/20060119/REVIEWS/60117003/1023.

Egan, Charles E. 1943. "Army seizes poultry in trucks to meet needs, end black market." *New York Times*, July 22, p. 1, 13.

Elliott, Carl. 2004. *Better than well: American medicine meets the American dream*. New York: Norton.

Engdahl, William. 2005. "Bird flu & chicken factory farms: Profit bonanza for U.S. agribusiness." *Scoop Independent News*, November 29, http://www.scoop.co.nz/stories/print.html?path=HL0511/S00351.htm.

Eriksson, Jonas, Greger Larson, Ulrika Gunnarsson, Bertrand Bed'hom, Michele Tixier-Boichard, Lina Stromstedt, Dominic Wright, Annemieke Jungerius, Addie Vereijiken, Ettore Randi, et al. 2008. "Identification of the *Yellow Skin* Gene Reveals a Hybrid Origin of the Domestic Chicken." *PLoS Genetics* 4, no. 2 (February 28): e1000010. doi:10. 1371/journal.pgen.1000010.

Fiedler, Leslie. 1992. *Love and death in the American novel.* New York: Anchor Books. (Orig. pub. 1960.)

Fiege, John, dir. 2007. *Mississippi chicken* [film]. Written by Anita Grabowski and John Fiege. Produced by John Fiege, Anita Grabowski, and Victor Moyers.

Fine, Robert. 2003. "Taking the 'ism' out of cosmpolitanism: An essay in reconstruction." *European Journal of Social Theory* 6, no. 4: 451–470.

Fink, Deborah. 1987. *Open country, Iowa: Rural women, tradition and change.* Albany: State University of New York Press.

Fitzgerald, Deborah. 2003. *Every farm a factory: The industrial ideal in American agriculture.* New Haven and London: Yale University Press.

Fleck, Ludwik. 1979. *The genesis and development of a scientific fact.* Chicago: University of Chicago Press. (Orig. pub. 1935.)

Foner, Philip S., ed. 1950–1975. *The life and writings of Frederick Douglass.* 5 vols. New York: International Publishers.

Franklin, H. Bruce. 1982. "America as science fiction: 1939. *Science Fiction Studies* 9, no. 1 (March): 38–50.

Franklin, Sarah. 2007. *Dolly mixtures: The remaking of genealogy.* Durham, NC: Duke University Press.

Franklin, Sarah, Celia Lury, and Jackie Stacey. 2000. *Global nature, global culture.* London: Sage Press.

Furst, Randy. 2008. "'Obama bucks' illustration was created by a Minnesota DFLer." October 22. *Star Tribune.* http://www.startribune.com/politics/32703279.html?elr=KArks LckD8EQDUoaEyqyP4o:DW3ckUiD3aPc:_Yyc:aULPQL7PQLancho7DiUI.

Furuti, Carlos A. 2009. "Pseudocylindrical projections," August 31. http://www.progonos. com/furuti/MapProj/Dither/ProjPCyl/projPCyl.html.

Gaines, Jane. 2001. *Fire and desire: Mixed-race movies in the silent era.* Chicago: University of Chicago Press.

Galton, Francis. 1904. Letter to William Bateson, June 12. Eugenics Archive, University College London, GP, 245/3. Available at Dolan DNA Learning Center Web site, http://www. eugenicsarchive.org/eugenics/image_header.pl?id=2201&pr.

GenomeWeb staff. 2008. "Chicken DNA fails to support early Polynesian–South American trade theory." *GenomeWeb,* July 31. http://www.genomeweb.com/chicken-dna-fails-support-early-polynesian-south-american-trade-theory.

Gibson-Graham, J. K. 2006. *The end of capitalism (as we knew it): A feminist critique.* Minneapolis: University of Minnesota Press.

Gifford, Terry. 2000. "Pastoral, anti-pastoral, post-pastoral." In *The green studies reader,* ed. Jonathan Bate Laurence Coupe, 219–222. London: Routledge.

Gilchrist, Mary J., Christina Greko, David B. Wallinga, George W. Beran, David G. Riley, and Peter S. Thorne. 2007. "The potential role of concentrated animal feeding operations in infectious disease epidemics and antibiotic resistance." *Environmental Health Perspectives* 115, no. 2 (February): 313–316; doi:10.1289/ehp.8837.

Giles, James R., Lisa M. Olsen, and Patricia A. Johnson. 2006. "Characterization of ovarian surface epithelial cells from the hen: A unique model for ovarian cancer." *Experimental Biology and Medicine* 231:1718–1725.

Gongora, Jaime, Nicholas J. Rawlence, Victor A. Mobegi, Han Jainlin, Jose A. Alcalde, Jose T. Matus, Olivier Hanotte, Chris Moran, Jeremy J. Austin, Sean Ulm, et al. 2008. "Indo-European and Asian origins for Chilean and Pacific chickens revealed by mtDNA."

Proceedings of the National Academy of Sciences 105, no. 30 (July 19): 10308–10313. http://www.pnas.org/content/105/30/10308.

Gonzalez, Ed. 2003. "Film review: *Freaks*." *Slant Magazine* (October 29). http://www.slant-magazine.com/film/review/freaks/795.

Greger, Michael. 2006. *Bird flu: A virus of our own hatching.* New York: Lantern Books.

Guèye, E. F. 2005. "Gender aspects in family poultry management systems in developing countries." *World's Poultry Science Journal* 61 (March): 39–46.

Gurdon, Martin. 2004. *Hen and the art of chicken maintenance: Reflections on raising chickens.* Guilford, CT: The Lyons Press.

Hanh, Thich Nhat. 1998. *The heart of the buddha's teaching.* New York: Broadway Books.

Hanke, O. A. 1950. "Now let's develop the 'egg-of-tomorrow.'" *Poultry Tribune* 82:82. American Poultry Historical Society Personal Collection of O. August Hanke, Box 1 of 2, National Agricultural Library, Beltsville, MD.

Hanke, Oscar August, John L. Skinner, and James Harold Florea. 1974. *American poultry history, 1823–1973.* Madison, WI: American Printing and Publishing.

Haraway, Donna J. 2008. *When species meet.* Minneapolis: University of Minnesota Press.

Hart, John Fraser. 1998. *The rural landscape.* Baltimore, MD: Johns Hopkins University Press.

Hartman, Zenkai Blanche. 2001. "Beginner's mind." http://www.intrex.net/chzg/hartman4.htm.

Hoggart, Richard. 1998. *The uses of literacy.* Edison, NJ: Transaction Publishers. (Orig. pub. 1957.)

Horowitz, Roger. 2006. *Putting meat on the American table: Taste, technology, transformation.* Baltimore, MD: Johns Hopkins University Press.

Howe, Lucien. 1918. "The relation of hereditary eye defects to genetics and eugenics." *Journal of the American Medical Association* 70, no. 26 (June 29): 1994–1999.

Hughes, Bill. 1999. "The constitution of impairment: Modernity and the aesthetic of oppression." *Disability & Society* 14, no. 2: 155–172.

———. 2000. "Medicine and the aesthetic invalidation of disabled people." *Disability & Society* 15, no. 4: 559–569.

———. 2002. "Disability and the body." In *Disability studies today*, ed. Colin Barnes, Mike Oliver, and Len Barton, 38–76. London: Polity.

Huxley, Julian, prod. 1934. *The private life of the gannets.* London: London Film Productions.

International Chicken Genome Sequencing Consortium (ICGSC). 2004. "Sequence and comparative analysis of the chicken genome provide unique perspectives on vertebrate evolution." *Nature* 432 (December 9): 695–777.

James, Frank. 2009. "Flu vaccine's egg-free future." *The Two-Way: NPR's News Blog*, October 27. http://www.npr.org/blogs/thetwo-way/2009/10/flu_vaccines_eggfree_future.html.

Jameson, Fredric. 1983. "Postmodern and consumer culture." In *The anti-aesthetic: Essays on postmodern culture*, ed. Hal Foster, 111–125. London: New Press.

Jelenk, Richard. 2005. "The contribution of new findings and ideas to the old principles of teratology." *Reproductive Toxicology* 20:295–300.

Johnson, Jenell. 2008. "Echoes of the soul: A rhetorical history of lobotomy." Ph.D. diss., The Pennsylvania State University, University Park.

Jones, Andy. 2000. "Reading August Sander's archive." *Oxford Art Journal* 23, no. 1: 1–22.

Jones, Roy E. 1944. *A basic chicken guide for the small flock owner.* New York: William Morrow and Company.

Kains, M. G. 1910. *Profitable poultry production.* New York: Orange Judd Company.

Kant, Immanuel. 1917. *Perpetual peace: A philosophical sketch.* London: George Allen & Unwin Ltd. (Orig. pub. 1795.)

Kantor, Jodi. 2008. "Teaching law, testing ideas, Obama stood slightly apart." *New York Times*, July 30. http://www.nytimes.com/2008/07/30/us/politics/30law.html?scp=1&sq =%93Teaching%20Law,%20Testing%20Ideas,%20obama%20Stood%20Slightly%20 Apart,%94&st=cse.

Keilholz, F. J. 1946. "Preview of tomorrow's chicken." *Country Gentleman* 116, no.1: 55–56.

Kelly, John. 2008. "Inaugural balls: For the birds?" *Washington Post* online, November 7. http://voices.washingtonpost.com/commons/2008/11/inaugural_balls_for_the_birds. html.

Kennedy, Tanya Ann. 2006. "The secret properties of Southern regionalism: Gender and agrarianism in Glasgow's barren ground." *Southern Literary Journal* 38, no. 2: 40–63.

Kevles, Daniel. 1985. *In the name of eugenics: Genetics and the uses of human heredity.* Cambridge, MA: Harvard University Press.

Kilpatrick, David. 2004–2005. "The pathological passion of Dario D'Ambrosi." *The Brooklyn Rail.* December–January. http://www.brooklynrail.org/2005/01/theater/the-pathological-passion-of-dario.

King, Lovalerie. 2007. *Race, theft and ethics: Property matters in African American literature.* Baton Rouge: Louisiana State University Press.

Kingsolver, Barbara, with Steven L. Hopp and Camille Kingsolver. 2007. *Animal, vegetable, miracle: A year of food life.* New York: HarperCollins.

Kirby, David. 2009. "Swine flu outbreak—Nature biting back at industrial animal production?" *Huffington Post*, April 25. http://www.huffingtonpost.com/david-kirby/swine-flu-outbreak—nat_b_191408.html.

Kirby, Vicki. 1999. "Human nature." *Australian Feminist Studies* 14, no. 29: 19–29.

Kleingeld, Pauline, and Eric Brown. 2009. "Cosmopolitanism," *The Stanford Encyclopedia of Philosophy* (Summer), ed. Edward N. Zalta. http://plato.stanford.edu/archives/sum 2009/entries/cosmopolitanism/.

Koen, J. S. 1919. "A practical method for field diagnosis of swine disease." *American Journal of Veterinary Medicine* 14:468.

Kolodny, Annette. 1975. *The lay of the land: Metaphor as experience and history in American life and letters.* Chapel Hill: University of North Carolina Press.

Komzak, Jiri, and Marc Eisenstadt. 2001. *Visualization of entry distribution in very large scale spatial and geographic information systems* (Milton Keynes, U.K.: Knowledge Media Institute, The Open University, June). http://kmi.open.ac.uk/publications/pdf/kmi-01-19.pdf.

Kornreich, Lauren. 2008. "Obama: New dog could be 'mutt' like me." CNN Politics, Political Ticker, November 7. http://politicalticker.blogs.cnn.com/2008/11/07/obama-new-dog-could-be-mutt-like-me/.

Kowalsky, Meaghan. 2005. "Review: *Disability and social policy in Britain since 1750: A history of exclusion.*" Review no. 453. *Institute of Historical Research/The National Centre for History.* http://www.history.ac.uk/reviews/review/453.

Kramer, Stanley, dir. 1951. *Judgment at Nuremburg.* Writ. Abby Mann. Roxlom Films, Inc.

Kuczewski, Mark. 2001. "Disability: An agenda for bioethics." *American Journal of Bioethics* 1, no. 3 (Summer): 36–44.

Lance, Stephen, dir. 2007. *Yolk.* Damon Escott, prod. Head Pictures.

Landecker, Hannah. 2007. *Culturing life: How cells became technologies.* Cambridge, MA: Harvard University Press.

Laughlin, Harry H. 1907. First letter to C. Davenport. February 25. American Philosophical Society. American Philosophical Society, Dav, B:D27,Ser 2,CSH-ERO. Available at Dolan DNA Learning Center Web site, http://www.eugenicsarchive.org/eugenics/view_ image.pl?id=506.

Lawrence, D. H. 1964. *Studies in classic American literature.* London: Heinemann. (Orig. pub. 1923.)

Lee, Andy, and Pat Foreman. 2006. *Chicken tractor: The permaculture guide to happy hens and healthy soil.* Buena Vista, VA: Good Earth Publications, LLC.

Le Guin, Ursula. 1996. "The carrier bag theory in fiction." In *The Ecocriticism reader: Landmarks in literary ecology,* ed. C. Glotfelty and H. Fromm, 149–154. Athens: University of Georgia Press.

Leonard, Andrew. 2009. "Swine flu on the automated pig farm." *Salon.* September 4. http://www.salon.com/technology/how_the_world_works/2009/09/04/swine_flu_pig_farm_update.

Levidow, Les. 1996. "Simulating mother nature, industrializing agriculture." In *FutureNatural: Nature/science/culture,* ed. George Robertson, Melinda Mash, Lisa Tickner, John Bird, Barry Curtis, and Tim Putnam, 55–71. London: Routledge.

Levister, Chris. 2008. "KFC demands cease and desist of racist 'Obama Bucks.'" *Inland Empire,* October 23. Posted by *Black Voice News.* http://blackvoicenews.com/news/42756-kfc-demands-cease-and-desist-of-racist-obama-bucks-.html.

Lewis, Harry R. 1919. *Making money from hens.* Philadelphia: J. B. Lippincott.

Lippit, Akira Mizuta. 2002. " . . . From wild technology to electric animal." In *Representing animals,* ed. Nigel Rothfels, 119–138. Bloomington: Indiana University Press.

Lipscomb, Hester J., Robin Argue, Mary Anne McDonald, John M. Dement, Carol A. Epling, Tamara James, Steve Wing, and Dana Loomis. 2005. "Exploration of work and health disparities among black women employed in poultry processing in the rural South." *Environmental Health Perspectives* 113, no. 12: 1833–1840.

Loboa, Linda, and Katherine Meyer. 2001. "The great agricultural transition: Crisis, change, and social consequences of twentieth century U.S. farming." *Annual Review of Sociology* 27: 103–124.

Locke, John. 1689. *Two treatises of government.* Edited by Peter Laslett. Cambridge: Cambridge University Press.

Lombardo, Paul A. 1985. "Three generations, no imbeciles: New light on *Buck v. Bell.*" *New York University Law Review* 60, no. 30: 30–62.

———. 2008. *Three generations no imbeciles: Eugenics, the Supreme Court, and Buck v. Bell.* Baltimore, MD: Johns Hopkins University Press.

Longmore, Paul K. 1995. "The second phase: From disability rights to disability culture." *Disability Rag & Resource* (September–October). Reprinted by Independent Living Institute. http://www.independentliving.org/docs3/longm95.html.

Lopez, I. Haney. 1996. *White by law: The legal construction of race.* New York: New York University Press.

Lott, Eric. 1992. "Love and theft: The racial unconscious of blackface minstrelsy." *Representations* 39 (Summer): 23–50.

———. 1993. *Love and theft: Blackface minstrelsy and the American working class.* New York: Oxford University Press.

Lovgren, Stefan. 2004. "Is Asian bird flu the next pandemic?" *National Geographic News,* December 7. http://news.nationalgeographic.com/news/2004/12/1207_041207_birdflu.html.

Luce, Nancy. 1875. "Poor little hearts." In *A complete edition of the works of Nancy Luce, of West Tisbury, Duke's County, Mass., containing God's words—sickness—poor little hearts—milk—no comfort—prayers—our savior's golden rule—hen's names, etc.* New Bedford: Mercury Job Press.

Lyon, Janet. 1999. *Manifestoes: Provocations of the modern.* Ithaca, NY: Cornell University Press.

MacDonald, Betty. 1945. *The egg and I.* New York: Harper & Row.

Marion, Frank J., and Wallace McCutcheon. 1904. "The chicken thief." Advertising Bulletin no. 39, December 27. In *Biograph bulletins 1896–1908*. Ed. Bebe Bergsten, 2. Los Angeles: Artisan Press.

Martin, Douglas. 2006. "Robert C. Baker, who reshaped chicken dinner, dies at 84." *New York Times*, March 16. http://www.nytimes.com/2006/03/16/nyregion/16baker.html?3i=5070&en=a5e6b31f0da9.

Marvell, Andrew. "The garden." In *Elizabethan and seventeenth-century lyrics*, ed. Matthew W. Black, 373–375. Philadelphia: J. B. Lippincott and Company, 1938.

Marx, Leo. 1974. *The machine in the garden: Technology and the pastoral ideal in America.* Oxford: Oxford University Press. (Orig. pub. 1964.)

McKenzie, J. 1967. "The chick embryo grown *in vitro.*" In *Teratology: Proceedings of a symposium organized by the Italian Society of Experimental Teratology*, October 21–22, ed. Aldo Bertelli, 43–54. Rome: Excerpta Medica Foundation.

McKibben, Bill. 2007. *Deep economy: The wealth of communities and the durable future.* New York: Henry Holt.

McNeil, Donald G., Jr. 2006. "From the chickens' perspective, the sky really is falling." *New York Times*, March 28, http://www.nytimes.com/2006/03/28/science/28bird.html?ex=1185422400&en=f6a3406d99cb0720&ei=5070.

Midkiff, Ken. 2004. *The meat you eat: How corporate farming has endangered America's food supply.* New York: St. Martin's.

Milway, Katie Smith. 2008. *One hen: How one small loan made a big difference.* Tonawanda, New York: Kids Can Press.

Mitman, Gregg. 1999. *Reel nature: America's romance with wildlife on film.* Cambridge, MA: Harvard University Press.

Mosher, Loren. 1982. "Italy's revolutionary mental health law: An assessment." *American Journal of Psychiatry* 139, no. 2 (February): 199–203.

Najafi, Hassan. 1983. "Dr. Alexis Carrel and tissue culture." *Journal of the American Medical Association* 250, no. 8 (August 26): 1086–1089.

Needham, Joseph, with Arthur Hughes. 1959. *A history of embryology.* 2nd ed., rev. New York: Abelard-Schuman.

Nelson, Pamela B. 2001. "Toys as history: Ethnic images and cultural change." http://www.ferris.edu/jimcrow/links/toys/.

Nicholson, Karl G. 2009. "Influenza and vaccine development: A continued battle." *Expert Reviews Vaccines* 8, no. 4: 373–374. http://www.expert-reviews.com/doi/pdf/10.1586/erv.09.17.

Niver, Kemp R. 1971. *Biograph bulletins 1896–1908.* Ed. Bebe Bergsten. Los Angeles: Artisan Press.

Norden, Martin F., and Madeline A. Cahill. 1998. "Violence, women, and disability in Tod Browning's 'Freaks' and 'the devil doll.'" *Journal of Popular Film & Television* 26, no. 2 (Summer): 86–94.

Obama, Barack. 2004. *Dreams from my father: A story of race and inheritance.* New York: Three Rivers Press.

OED. 1971. *The compact edition of the Oxford English dictionary.* Vols. A–O. 2 vols. Oxford: Oxford University Press.

Oppenheimer, Jane M. 1968. "Some historical relationships between teratology and experimental embryology." *Bulletin of the History of Medicine* 42, no. 2 (March–April): 145–159.

Orenstein, Peggy. 2010. "The femivore's dilemma," *New York Times Sunday Magazine*, March 11, p. 11.

Ozeki, Ruth. 1998. *My year of meats.* New York: Penguin.

———. 2006. "The death of the last white male." *Configurations* 14, no. 1–2 (Winter–Spring): 61–68.

Pagliuso, Jean. 2006. Interview by Susan Squier. New York City, NY. December 18.

Palmer, A. K. 1967. "The relationship between screening tests for drug safety and other teratological investigations." In *Teratology: Proceedings of a symposium organized by the Italian Society of Experimental Teratology*, 55–72. Amsterdam: Excerpta Medica Foundation.

Paul, Johannes, and William Wyndham. 2007. Interview by Susan Squier. Omlet headquarters, Wardington, Oxfordshire, UK. November 7.

Pauly, Philip. 1987. *Jacques Loeb and the engineering ideal in biology.* Oxford: Oxford University Press.

———. 2007. Fruits and plains: The horticultural transformation of America. Cambridge, MA: Harvard University Press.

Pearce, Richard, dir. 2006. *Fatal contact: Bird flu in America.* Writ. Ron McGee. ABC, May 9.

Percy, Pam. 2002. *The complete chicken.* Minneapolis: Voyageur Press.

———. 2006. *The field guide to chickens.* Minneapolis: Voyageur Press.

Perelman, S. J. 1952. *Chicken inspector no. 23.* New York: Simon & Schuster.

Pew Commission on Industrial Farm Animal Production. 2008. "Putting meat on the table: Industrial farm animal production in America." August 3. http://www.ncifap.org/bin/e/j/PCIFAPFin.pdf.

Phillips, Mike. 2008. "Our Obama moment." *Guardian*, November 6. http://www.guardian.co.uk/commentisfree/2008/nov/05/uselections2008-race.

Platt, Frank L. 1939. "How will the war affect the U.S. poultry industry?" Condensed from the *American Poultry Journal*, *National Poultry Digest* 1, no. 10 (October): 577–579.

Pollan, Michael. 2006. *The omnivore's dilemma: A natural history of four meals.* New York: Penguin.

Proctor, Robert. 2008. "Agnotology: A missing term to describe the cultural production of ignorance (and its study)." In *Agnotology: The making and unmaking of ignorance*, ed. Robert Proctor and Londa Schiebinger, 1–36. Stanford, CA: Stanford University Press.

Rankin, Katherine N. 2001. "Governing development: Neoliberalism, microcredit, and rational economic woman." *Economy and Society* 30, no. 1: 18–37.

Rasmussen, W. D. 1989. *Taking the university to the people: Seventy-five years of cooperative extension.* Ames: Iowa State University Press.

Redgrove, Herbert Stanley. 1920. *Bygone beliefs: Being a series of excursions in the byways of thought.* London: William Rider & Son, Ltd. http://onlinebooks.library.upenn.edu/webbin/book/lookupname?key=Redgrove%2C%20H.%20Stanley%20(Herbert%20Stanley)%2C%201887–1943.

Richards, Evelleen. 1994. "A political anatomy of monsters, hopeful and otherwise." *Isis* 85:377–411.

Ridley, Mark. 2006. *How to read Darwin.* New York: W. W. Norton.

Ritvo, Harriet. 1997. *The platypus and the mermaid: And other figments of the classifying imagination.* Cambridge, MA: Harvard University Press.

Roberts, Dorothy. 1998. *Killing the black body: Race, reproduction, and the meaning of liberty.* New York: Vintage Books.

———. 2006. "Legal constraints on the use of race in biomedical research: Towards a social justice framework." *Journal of Law, Medicine, and Ethics* 34, no. 3 (Fall): 526–534.

Robinson, John H. 1912a. *Principles and practice of poultry culture.* Boston: Athenaeum/Ginn and Company.

———. 1912b. *Standard poultry for exhibition: A complete manual of the methods of expert exhibitors in growing, selecting, conditioning, training and showing poultry—Fully describing fitting processes and exposing faking practices—Briefly explaining judging for the amateur and furnishing the student of judging an exhaustive analysis of the history, philosophy and merits of comparison and score-card systems.* Quincy, IL: Reliable Poultry Journal Publishing Co.

Rous, Peyton, and James D. Murphy. 1912. "The nature of the filterable agent causing a sarcoma of the fowl." *Journal of the American Medical Association* 58, no. 22: 1938.

Rous, Peyton, James D. Murphy, and W. H. Tytler. 1912. "Transplantable tumors of the fowl: A neglected material for cancer research." *Journal of the American Medical Association* 58, no. 22: 1682–1683.

Ruvinsky, Jessica. 2007. "Could Autism Be the Next Stage of Human Evolution?" November 6. *Discover Magazine, Blogs/Discoblog* http://blogs.discovermagazine.com/discoblog/2007/11/06/could-autism-be-the-next-stage-of-human-evolution/.

Sachs, Carolyn. 1996. *Gendered fields: Rural women, agriculture, and environment.* Boulder, CO: Westview Press.

Salatin, Joel. 1993. *Pastured poultry profits.* Swope, VA: Polyface.

Sawyer, Gordon. 1971. *The agribusiness poultry industry: A history of its development.* New York: Exposition Press.

Sayadow, Venerable Mahasi. 1999. *A discourse on dependent origination.* Trans. U Aye Maung. Bangkok: Buddhadamma Foundation. Online at http://www.vipassana.com.my/eBooks/DepOrigin.pdf.

Sayer, Karen. 2007. "'Let nature be your teacher': Tegetmeier's distinctive ornithological studies." *Victorian Literature and Culture* 35:589–605.

Schechner, Richard. 2003. *Performance theory.* New York: Routledge.

Schell, Orville. 2002. "Gross national happiness," PBS *Frontline,* May. http://www.pbs.org/frontlineworld/stories/bhutan/gnh.html. (Originally published in *Red Herring* January 15, 2002.)

Schlosser, Eric. 2001. *Fast food nation: The dark side of the all-American meal.* New York: Harper Perennial.

Schmidt, Charles W. 2009. "Swine CAFOs and Novel H1N1 flu: Separating facts from fears." *Environmental Health Perspectives* 117, no. 9 (September 1): A394–A401. doi:10.1289/ehp.117-a394.

Schmutz, Jeremy, and Jane Grimwood. 2004. "Fowl sequence." *Nature* 432 (December 9): 679–680.

Schoor, Frank von de, and Jo Coucke. 2008. *Koen Vanmechelen: The chicken's appeal,* Catalogue. Nijmegen: Museum Het Valkhof.

Shakespeare, Tom. 2005. "Review essay: Disability studies today and tomorrow." *Sociology of Health and Illness* 27, no. 1: 138–148.

Siebers, Tobin. 2008. *Disability theory.* Ann Arbor: University of Michigan Press.

Siegel, Taggart, dir. 2005. *The real dirt on Farmer John.* Taggart Siegel and Terri Lang, prod. Independent Lens.

Simons, Barbara, and Wouter Keirse. 2003. *Koen Vanmechelen cosmopolitan chicken project.* Amsterdam: Ludion Ghent-Amsterdam.

Skinner, John L. 1974. *American poultry history 1823–1973.* Madison, WI: American Printing and Publishing.

Slaff, Jonathan. 2007. "'Days of Antonio' at La Mama." *LoHo 10002,* http://www.loho10002.com/wordpress/?p=1163.

Smith, Adam. 1904. *An inquiry into the nature and causes of the wealth of nations.* London, Methuen. (Orig. pub. 1776.) The Library of Economics and Liberty Web site. http://www.econlib.org/library/Smith/smWN.html.

———. 1976. *The theory of moral sentiments.* London: A. Millar. (Orig. pub. 1759.)

Smith, Elmer Boyd. 1910. *Chicken world*. New York: Knickerbocker Press, 1910.

Smith, Henry Nash. 1970. *Virgin land*. Cambridge, MA: Harvard University Press. (Orig. pub. 1950.)

Smith, William. 1875. *A dictionary of Greek and Roman antiquities*. London: John Murray. http://penelope.uchicago.edu/Thayer/E/Roman/Texts/secondary/SMIGRA*/Haruspices.html.

Snow, C. P. 1933. *New lives for old*. London: Camelot.

———. 1993. *The two cultures*. Introduction by Stefan Collini. Cambridge: Cambridge University Press. (Orig. pub. 1959.)

Snyder, Sharon L., and David T. Mitchell. 2001. "Re-engaging the body: Disability studies and the resistance to embodiment." *Public Culture* 13, no. 3: 367–389.

Soper, Kate. 1996. "Nature/'nature.'" In *FutureNatural: Nature/science/culture*, ed. George Robertson, Melinda Mash, Lisa Tickner, John Bird, Barry Curtis, and Tim Putnam, 22–34. London: Routledge.

Squier, Susan M. 1994. *Babies in bottles: Twentieth-century visions of reproductive technology*. New Brunswick, NJ: Rutgers University Press.

———. 2004. *Liminal lives: Imagining the human at the frontiers of biomedicine*. Durham, NC: Duke University Press.

———. 2006. "Chicken auguries." *Configurations* 14, no. 1: 69–86.

Steck, Franz, and H. U. Haberstich. 1976. "Marek's disease in chickens: Development of viral antigen in feather follicles and of circulating antibodies." *Infection and Immunity* 13, no. 4: 1037–1045.

Stern, Claudio D. 2002. "Induction and initial patterning of the nervous system: The chick embryo enters the scene." *Current Opinion in Genetics & Development* 12:447–451.

———. 2005. The chick: A great model system becomes even greater." *Developmental Cell* 8:9–17.

———. 2007. Interview by Susan Squier. University College, London. November 6.

Stiker, Henri-Jacques. 1999. *A history of disability*. Trans. William Sayers. Ann Arbor: University of Michigan Press.

Stivale, Charles J. 2007. Summary of *l'abecedaire de Gilles Deleuze, avec Claire Parnet*, http://www.langlab.wayne.edu/CStivale/D-G/ABCs.html.

Stone, Marvin J. 2003. "History of the Baylor Charles A. Sammons Cancer Center." *Baylor University Medical Center Proceedings* 16, no. 1: 30–58.

Storey, Alice A., Jose Miguel Ramirez, Daniel Quiroz, David V. Burley, David J. Addison, Richard Walter, Atholl J. Anderson, Terry L. Hunt, J. Stephen Athens, Leon Huynen, et al. 2007. "Radiocarbon and DNA evidence for a pre-Columbian introduction of Polynesian chickens to Chile." *Proceedings of the National Academy of the Sciences* 104, no. 25 (June 19): 10335–10339. http://www.pnas.org/content/104/25/10335.abstract.

Strode, A. 1925. "Letter to H. Laughlin, requesting a deposition on hereditary feeblemindedness for the trial of Carrie Buck in Amherst, Virginia." September 30. University of Albany, SUNY, Estabrook, SPE, SMX 80.9 Bx 1 folder 1–40. Available at http://www.eugenicsarchive.org.

Striffler, Steve. 2005. *Chicken: The dangerous transformation of America's favorite food*. New Haven, CT: Yale University Press.

Strunk, William, Jr., and E. B. White. 1962. *The elements of style*. New York: Macmillan.

subRosa. 1995. "Cultures of eugenics," pamphlet, rev. 3rd ed., unlimited ed. (Produced for Express Choice).

Sugden, Robert. 2002. "Beyond sympathy and empathy: Adam Smith's concept of fellow-feeling." *Economics and Philosophy* 18:63–87.

Shunryu Suzuki. 1970. *Zen mind, beginner's mind*. New York: Weatherhill.

Taubenberger, Jeffrey K., Ann H. Reid, Raina M. Lourens, Ruixue Wang, Guozhong Jin, and Thomas G. Fanning. 2005. "Characterization of the 1918 influenza virus polymerase genes." *Nature* 437 (October 6): 889–893.

Tegetmeier, W. B. 1890. "On the principal modern breeds of the domestic fowl." *Ibis* 32, no. 3 (July): 304–327.

Thaxton, Y. Vizzier, J. A. Cason, N. A. Cox, S. E. Morris, and J. P. Thaxton. 2003. "The decline of academic poultry science in the United States of America." *World's Poultry Science Journal* 59, no. 3 (September): 303–313.

Thompson, Paul B. 2007. "Agriculture and working-class political culture: A lesson from *The grapes of wrath*." *Agriculture and Human Values* 24:165–177.

Thomson, Rosemarie Garland, ed. 1996. *Freakery: Cultural spectacles of the extraordinary body*. New York: New York University Press.

———. 1997. *Extraordinary bodies: Figuring physical disability in American culture and literature*. New York: Columbia University Press.

Thurtle, Philip. 2007. *The emergence of genetic rationality: Space, time, and information in American biological science, 1870–1920*. Seattle: University of Washington Press.

Tilman, David, Kenneth G. Cassman, Pamela A. Batson, Rosamond Naylor, and Stephen Polasky. 2002. "Agricultural sustainability and intensive production practices." *Nature* 418 (August 8): 671–677.

Tumpey, Terrence M., Christopher F. Basler, Patricia V. Aguilar, Hui Zeng, Alicia Solórzano, David E. Swayne, Nancy J. Cox, Jacqueline M. Katz, Jeffrey K. Taubenberger, Peter Palese, et al. 2005. "Characterization of the reconstructed 1918 Spanish influenza pandemic virus," *Science* 310, no. 5745 (October 7): 77–80.

Turner, Victor. 1977. "Frame, flow, and reflection: Ritual and drama as public liminality." In *Performance in postmodern culture*, ed. Michel Benamou and Charles Carmello, 33–55. Madison, WI: Coda Press.

U.S. Food and Drug Administration (FDA). 2009. "FDA approves vaccines for 2009 H1N1 influenza virus." FDA News Release, September 15. http://www.fda.gov/NewsEvents/Newsroom/PressAnnouncements/2009/ucm182399.htm.

U.S. Government Accountability Office (GAO). 2005. "Workplace safety and health: Safety in the meat and poultry industry, while improving, could be further strengthened." January. http://www.gao.gov/new.items/d0596.pdf.

Van Horn, Beth E., Constance A. Flanagan, and Joan S. Thomson. 1998. "The first fifty years of the 4-H program." *Journal of Extension* 36, no. 6 (December). http://www.joe.org/joe/1998december/comm2.php.

Vanmechelen, Koen. 2007. "Medusa, the accident: Chronicles of the cosmopolitan chicken." Stekene, Belgium: Verbeke Foundation, June.

———. 2008. *Het appel van de kip: The chicken's appeal*. Nijmegen: Museum Het Valkhof.

Varadarajan, Siddharth. 2005. "India tertia and the mapping of the colonial imaginary." *Reality, One Bite at a Time: India, Asia, and the World* [blog], April 6. http://svaradarajan.blogspot.com/2005/04/india-tertia-and-mapping-of-colonial.html.

Vujakovic, Peter. 2003. "Damn or be damned: Arno Peters and the struggle for the 'new cartography.'" *Cartographic Journal* 40, no. 1 (June): 61–67.

Waddington, Conrad H. 1932. "Experiments on the development of chick and duck embryos, cultivated in vitro." *Philosophical Transactions of the Royal Society of London, Series B, Containing Papers of a Biological Character* 221:179–230.

———. 1952. *The epigenetics of birds*. Cambridge: Cambridge University Press.

Wald, Priscilla. 2008. *Contagious: Cultures, carriers, and the outbreak narrative*. Durham, NC: Duke University Press.

Waldby, Catherine. 2000. *The visible human project: Informatic bodies and posthuman medicine.* London: Routledge.

Wallace, Henry C. 1924. "A national agricultural program: A farm management problem." *Journal of Farm Economics* 6 (January): 1–7.

Warner, Melanie. 2005. "Sharpton joins with an animal rights group in calling for a boycott of KFC." *New York Times,* February 2, 2005. http://www.nytimes.com/2005/02/02/business/02chicken.html?

Washington, Booker T. 1901. *Up from slavery: An autobiography.* Garden City, NY: Doubleday. http://www.online-literature.com/booker-washington/up-from-slavery/.

Weinbaum, Alys. 2004. *Wayward reproductions.* Durham, NC: Duke University Press.

Weiss, Kenneth M., and Anne V. Buchanan. 2009. "The cooperative genome: Organisms as social contracts." *International Journal of Developmental Biology* 53:753–763.

West, Michael D. 1968. "Sherwood Anderson's triumph: 'The egg.'" *American Quarterly* 20, no. 4 (Winter): 675–693.

White, E. B. 1944. "Introduction." In *A basic chicken guide for the small flock owner,* by Roy E. Jones. New York: William Morrow and Company.

Wiggins, William H., Jr. 1988. "Boxing's Sambo twins: Racial stereotypes in Jack Johnson and Joe Louis newspaper cartoons, 1908 to 1938." *Journal of Sport History* 15, no. 3 (Winter): 242–254.

Wilkinson, Tom. 1951. "The egg used to come first but: Chicken-of-tomorrow changed poultry's face," *Arkansas Agriculturalist* 28:8; archived in the American Poultry Historical Society Personal Collection of Robert M. Smith, Box 2. National Agricultural Library, Beltsville, MD.

Williams, Linda. 2001. *Playing the race card: Melodramas of black and white from Uncle Tom to O. J. Simpson.* Princeton, NJ: Princeton University Press.

Williams, Raymond. 1973. *The country and the city.* Oxford: Oxford University Press.

———. 1983. *Keywords: A vocabulary of culture and society.* Rev. ed. New York: Oxford University Press.

———. 1989. "Culture is ordinary." In *Resources of hope: Culture, democracy, socialism,* 3–18. London: Verso.

Williams-Forson, Psyche. 2006. *Building houses out of chicken legs: Black women, food, and power.* Chapel Hill: University of North Carolina Press.

Wilmot, Sarah. 2007. "Introduction: Between the farm and the clinic; agriculture and reproductive technology in the twentieth century." *History and Philosophy of Science, Part C: Studies in History and Philosophy of Biology and the Biomedical Sciences* 38, no. 2: 302–315.

Wong, G.K.S., B. Liu, J. Wang, Y. Zhang, X. Yang, Z. J. Zhang, Q. S. Meng, et al. 2004. "A genetic variation map for chicken with 2.8 million single-nucleotide polymorphisms." *Nature* 432 (December 9): 717–722; doi:10.1038/nature03156.

World Health Organization (WHO). 2010. Table, "Cumulative number of confirmed human cases of avian influenza A/(H5N1) reported to WHO," January 28. http://www.who.int/csr/disease/avian_influenza/country/cases_table_2010_01_28/en/index.html.

Yablonsky, Lisa. 2006. "West egg story." August 15. http://www.artforum.com/diary/id=11472.

Young, Robert J. C. 1995. *Colonial desire: Hybridity in theory, culture and race.* New York: Routledge.

Youngquist, Paul. 2003. *Monstrosities: Bodies and British romanticism.* Minneapolis: University of Minnesota Press.

INDEX

Italicized page numbers refer to figures.

ABOUT THE AUTHOR

Susan Squier is Brill Professor of Women's Studies; English; and Science, Technology, and Society (STS), and acting director of STS (2010–2011) at the Pennsylvania State University. The recipient of the Graduate Teaching Award at Penn State University, she was codirector, with Anne Hunsaker Hawkins, of the National Endowment for the Humanities (NEH) Summer Institute in Medicine, Literature, and Culture at Penn State Hershey Medical Center, July 7–August 2, 2002. A scholar in residence at the Bellagio Study and Conference Center of the Rockefeller Foundation in 2001, she serves on the editorial board of the *Journal of the Medical Humanities* and as an executive board member and past president of the Society for Literature and Science.

She is the author of *Babies in Bottles: Twentieth-Century Visions of Reproductive Technology* (Rutgers University Press, 1994); *Liminal Lives: Imagining the Human at the Frontiers of Biomedicine* (Duke University Press, 2004); and *Virginia Woolf and London: The Sexual Politics of the City* (University of North Carolina Press, 1985); co-editor of *Playing Dolly: Technocultural Figurations, Fantasies, and Fictions of Assisted Reproduction* (Rutgers, 1999), and *Arms and the Woman: War, Gender, and Literary Representation* (University of North Carolina Press, 1989); and editor of *Communities of the Air: Radio Century, Radio Culture* (Duke University Press, 2003) and *Women Writers and the City: Essays in Feminist Literary Criticism* (University of Tennessee Press, 1984).

Squier and her husband live (with chickens) in Boalsburg, Pennsylvania, and (without chickens) in New York City.